Understanding
Hi-Fi Specifications

Understanding Hi-Fi Specifications

John Earl

Fountain Press

ARGUS BOOKS LTD
14 St James Road
Watford, Herts, England

Consultant Editor: Norman Stevens

Fountain Press
Argus Books Ltd
14 St James Road
Watford, Herts, England

ISBN 0 85242 582 1

First published 1978

Filmset by
The Humble Wordsmith Ltd, Tonbridge, Kent

Printed and bound in Great Britain by
A. Wheaton & Co. Ltd., Exeter

CONTENTS

PREFACE

THIS NEW BOOK is styled to complement the ever-growing series of books on hi-fi subjects under our *John Earl* name. It brings together all the major specifications of hi-fi equipment and separately investigates each parameter.

The plan has been to interpret the parameters in the least technical manner possible, to help towards an understanding of their meanings, to help with the choice of hi-fi equipment from manufacturers' specifications and from the tested parameters in hi-fi magazine reviews and to see how different makes and designs of hi-fi equipment compare technically relative to purchase price. It is known that undue simplification can seriously detract from the value of a book owing to technical ambiguity. We have been on guard to avoid this.

Although not a text book in the true sense of the word, it is directed as an authoritative 'text' to the hi-fi enthusiast, hi-fi dealer and his staff and to the student of matters hi-fi. The level (it is hoped!) is set so as not to insult the technical sensibility of our readership, and to this end we have avoided kindergarten colloquialisms.

We have found from our wide contacts with hi-fi people — both technical and not so technical — that nursery language is not only insulting to them but also that it fails to lead to a proper understanding of the principles involved. Although sometimes regarded as an art — and, more recently, as something of a 'witchcraft' — hi-fi is also an intrinsically exacting science which, for its full appreciation, must be treated as such. It is certainly not outside the ken of the average chap in the street to appreciate, if not fully to understand.

A good deal of the information presented in this book has been researched in the labs of *Gordon J. King (Enterprises) Limited,* to whom full acknowledgement is given, and a number of the parameters looked at reflect latter-day thinking and measuring techniques.

You will note in Chapter 8 that we fail to subscribe to the passing school of thought which belittles physical measurements and, particularly, that which places undue emphasis on solitary 'opinion' of a listener as the primary judgement of reproduction quality.

It is our considered judgement that if any item of hi-fi equipment is conclusively proved to be relatively 'poor' or relatively 'good' during audition, then there are immutable physical laws responsible, which must obviously be objectively detectable (by instrument tests). The excellent article by *John Gardner,* entitled Gizmology, and published on page 60 of *Practical Hi-F and Audio of September, 1977* exposes some interesting points in this respect, and is well worth reading, even if it means going to the local library to seek out the copy!

We are certainly not opposed to listening tests to supplement lab measurements; but these *must* be as well controlled and as scientific as the lab measurements themselves. Indeed, a full assessment of hi-fi equipment *demands* auditioning. Our lab has available a long-standing listening panel of ten to twelve people, all of whom have been 'weighted' and tested, with a large proportion concert goers on a regular basis or in other ways directly connected with live music.

Sound is rather like colour — not all people discern it the same way. Hence the danger of relying solely upon 'personal opinion' and the adjectives — sometimes invented ones — to describe this. Parameter measurements will always be essential for a proper evaluation of hi-fi equipment. The pinnacle, of course, is to seek out those measurements which have the greatest correlation with the listening experience, and this is just what has been happening in our lab over the last several years in conjunction with our listening panel.

1978 JOHN EARL

CHAPTER ONE

WHAT IS A SPECIFICATION?

MOST THINGS TECHNICAL are designed to a specification. An electric fire, for example, may have a specification reading '1kW at 240V'. There are two items involved here — one the input voltage and the other the power loading — and they are usually expressed separately, so giving the more detailed specification below:

Specification of Electric Fire
Input: 240V
Loading: 1,000W

Parameters

The term *parameter* is now commonly used to denote an item of a specification. Thus the *input* and *loading* of the specification above are parameters. Parameter is a mathematical term which serves to determine or define a quantity, point, line or figure. It is thus a fair term to use to signify an item of a specification, though one should not be unduly bothered about its mathematical overtone.

The chap who designed the electric fire above set out to ensure that when 240V mains is applied the element loads at 1kW. The element is wound with a special kind of resistance wire which is capable of operating at a red hot temperature without fusing. The resistance of such wire changes from its 'cold' value as it rises in temperature, increasing when its *temperature coefficient* is positive and decreasing when it is negative. One way that the designer can ensure that the loading is as per the requirement when 240V mains is applied across it is to look up the 'hot' resistance of the wire per unit length (or find this from the temperature coefficient) and then calculate the length of wire required to provide the loading.

Ohm's Law

All this may seem to be somewhat removed from hi-fi specs, but when we get to the next chapter (on amplifier specs) we shall see that it is highly pertinent! The loading or power dissipated in *watts* (W) is equal to the *voltage* (V) across the element *times* the *current* (I) flowing through it. It is much easier to appreciate this simple relationship by the use of symbols, thus: $W = V \times I$

Other simple relationships also come into the picture, such as where R is the resistance value in ohms (note that the *times* sign is generally omitted when symbols are used). By simple rearrangement this also yields:

$$I = \frac{V}{R} \text{ and } R = \frac{V}{I}$$

These are the relationships sorted out years ago by a Dr. Ohm and have since come to be known as Ohm's law expressions.

Now, going back to W = VI and taking due account of the Ohm's law expressions we can derive two additional expressions for power, which are:

W = IRI, simplifying to $W = I^2R$, and

$W = \frac{VV}{R}$, simplifying to $W = \frac{V^2}{R}$

Thus we have three expressions for power which are:

$$W = VI \quad \ldots\ldots\ldots\ldots\ldots\ldots (1)$$
$$W = I^2R \quad \ldots\ldots\ldots\ldots\ldots\ldots (2)$$
$$W = \frac{V^2}{R} \quad \ldots\ldots\ldots\ldots\ldots\ldots (3)$$

These involve the four *parameters* of power, voltage, current and resistance. To get to grips with the electric fire design, therefore, the designer uses these parameters together with those of temperature coefficient and resistance per unit length of element wire. There are others that he will also certainly have to consider, including those defining the current carrying capability and running temperature of the element wire, the insulation resistance to the accessible metal parts, earthing resistance, parameters of safety and so forth.

Growth of Parameters

We have shown that from the two simple parameters of the basic and possibly published specification of an electric fire, stem hosts of other parameters at the design stage. This is applicable to hi-fi kit as well as to electric fires, motor cars, lawn mowers — you name it! Joe Bloggs is not interested in the designer's problems. When he invests in an electric fire he is only interested in knowing that when he plugs it into his 240V mains supply it is going to load at 1kW and hence burn up one unit of expensive electricity each hour.

He will, though, want to be fully assured that it is safe, and the parameter that he looks to for this is that implicit in the stamp of approval given by a testing authority such as the British Standard Specification (BSI 415) or, more recently in the UK, the British Electromechanical Approvals Board (BEAB). Greater sensitivity to matters of safety is developing in hi-fi design and retail circles, and these points are stressed in the appropriate sections of the chapters that follow.

However, just to conclude our notes on the electric fire designer's problem, we recall that he wants the element to load to 1kW across 240V mains, so he can rearrange power expression (3) to give R from:

$$R = \frac{V^2}{W}$$

Substituting the figures we get:

$$R = \frac{240^2}{1{,}000}$$

which works out to 57·6 ohms.

He thus composes the element of suitable resistance wire which when running red hot has a value of 57·6 ohms.

The Customer's Viewpoint

A purchaser would be somewhat annoyed, to say the least, if he found that the 1kW fire was burning up 1·5kW/hours (i.e., 1·5 units each hour). Thus he has to take the published specification as gospel; that is, unless he decides to check the spec himself by timing the electric meter over one hour with only the fire of his electric system operating. Should he do this and find that 0·75 or 1·5 units are recorded, then either the spec is in error or the meter is inaccurate.

If the spec is wrong he has a good case to lodge a complaint with his local Trading Standards Department, getting them to tackle the supplier under the terms of the Trade Descriptions Act 1968. Before he does this, and hence takes issue with the supplier on false advertising and spec marking, he should first get his local Electricity Board to check the accuracy of his meter, for it may be that he is paying too much (or too little!) for the electricity he is burning.

Usually one would rarely go to all this trouble over the checking of an electric fire spec. The spec can never be one hundred per cent accurate, anyway, for if the mains voltage falls below the 240V nominal the power loading will also fall. The electric fire will give out less heat and consume less electricity. If the voltage drop is excessive, though, there may be a case for taking the Electricity Board to task!

Hi-Fi Specs

With hi-fi equipment it is different. There are many more spec parameters and hi-fi costs significantly more pounds sterling than electric fires. Moreover, it is far more specialised and one is greatly encouraged to buy on the strength of the published specification, manufacturers often competing with each other essentially in terms of the published specs.

It is very important, therefore, for users and potential purchasers of hi-fi equipment to know what the parameters of the specs are all about, and it is the primary objective of this new book to focus well and truly on specs, to detail and interpret the various parameters of all the items of a hi-fi system and ultimately to see whether it is possible from the parameters alone to determine how well a hi-fi system is likely to audition — to 'sound' in the listening room.

Use of Specifications

One objective of a spec is to allow a user to test for himself or have tested for him the separate parameters to ensure that they relate reasonably accurately to those published by the manufacturer, upon the strength of which the equipment was possibly purchased. The spec also gives a good clue as to the quality and design class of the equipment and its relationship to price.

It also tells whether an item of equipment such as a hi-fi amplifier or f.m. tuner will be suitable for the proposed application; for example, whether the amplifier is likely to result in sufficient sound intensity when driving loudspeakers of given sensitivity or efficiency in a room of specified volume and furnishings, and whether the tuner will be suitable for the reception of local or distant stations under the conditions prevailing at the reception site and when used with an aerial of given parameters.

Although it is unwise to choose hi-fi equipment on the specifications alone (it is highly desirable to judge by ear as well, and this applies particularly to loudspeakers and f.m. tuners), one cannot really make an intelligent choice without resource to the specs.

Hi-Fi Amplifier

Before making a choice of any hi-fi item, a number of questions need to be asked and answered. For example, when deciding on the purchase of an amplifer the questions go something like this: How large is the room? Is it heavily furnished with soft chairs and thick carpets or is it sparsely furnished with wooden floor and hard walls? Are large or relatively small loudspeakers to be used? Is hard rock, pop or classical music mostly to be indulged in?

All these things can be translated to spec parameters and it then becomes possible to choose an amplifier from a technical standpoint, if not from one of musicality, purely from the spec parameters.

Hi-Fi Tuner or Receiver

Will the tuner be required for the reception of stations in the a.m. bands as well as the mono and stereo ones in the f.m. band (remembering that only f.m. is capable of hi-fi quality and stereo). If the answer is 'yes', then what a.m. bands will be required (medium, long *and* medium *or* long, medium and short)? Most f.m./a.m. tuners and receivers have medium wave only. Long wave will now also be important owing to imminent a.m. wavelength changes by the BBC (November, 1978).

Will the near service area regional and 'local' f.m. stations only be used? If not, will the tuner be used for the reception of more distant stations, such as the *nearest* stereo station located beyond the accepted service area and hence in the fringe area or will it be used for more esoteric DX-ing (i.e., long-distance and freak reception trials)?

What are the local reception conditions like — i.e., is the reception site on the main road, partly or heavily screened or in open country? Is it possible to use an outside aerial or is it proposed to use an indoor or roof-space aerial?

Is the tuner or receiver to be used for the best possible hi-fi quality or will station seeking have greater priority — or will both be important? The answers to questions like these will determine the tuner parameters to look out for.

Loudspeakers

Having decided on the amplifier parameters, the plan then is to study the parameters of the loudspeakers to ensure compatibility. For example, are they capable of handling the input required for the output sound pressure in the listening room? Are they sufficiently sensitive (or efficient) to provide the required listening room peak sound pressure from the output power or voltage of the amplifier?

Is the listening room sufficiently large to do full justice to the extended bass output of the larger and hence more costly type of floor-standing loudspeakers (if not, then there is little future in spending quite a lot of extra cash in an endeavour to reproduce down to the lower octave — it will not be audible, see Chapter 6)? Are the cosmetics of the loudspeakers acceptable to the distaff side(!)?

The parameters alone will determine all these factors, but as loudspeakers are more 'personal' than the electronics hardware side of hi-fi the parameters of the spec alone certainly cannot tell how a pair of loudspeakers will audition (to you) in the listening room. A home trial is the answer — in your own room on your own equipment.

Ancillary Equipment

The spec parameters of ancillary equipment, such as the record and cassette (or tape) playing decks, headphones, f.m. aerial, etc., tell quite a lot about the technical performance but less about the aesthetics of the resulting sound and this — like loudspeakers — applies in particular to pickup cartridges and headphone sets.

Pickups especially have a characteristic 'sound' of their own. They are not absolutely neutral in this respect since they add their own bits of flavouring or coloration. In these areas choice is influenced more by quality compatibility than by technical compatibility. Pretty well any contemporary pickup cartridge will work with any contemporary amplifier or hi-fi receiver.

If you are spending £60 on an amplifier then you would choose a pickup which suits the quality of the amplifier at that low price — not one costing itself £60 to £80! Conversely, if you are running an amplifier or receiver of £300 or £400 quality it is hardly likely that you would partner it with a cartridge costing a fiver or so.

Pickups and loudspeakers tend to work hand-in-hand so far as the resulting

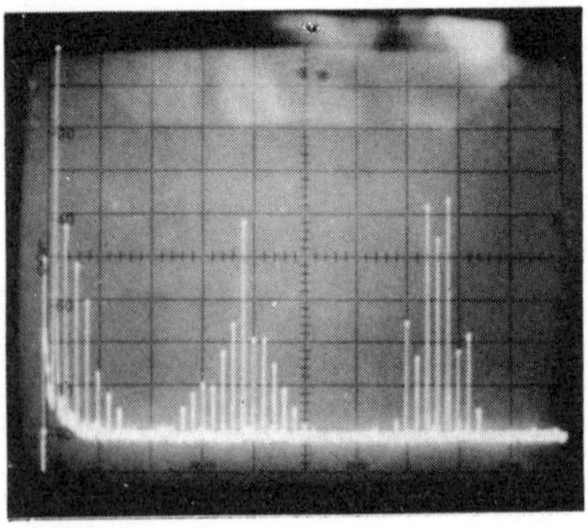

1.1(a)

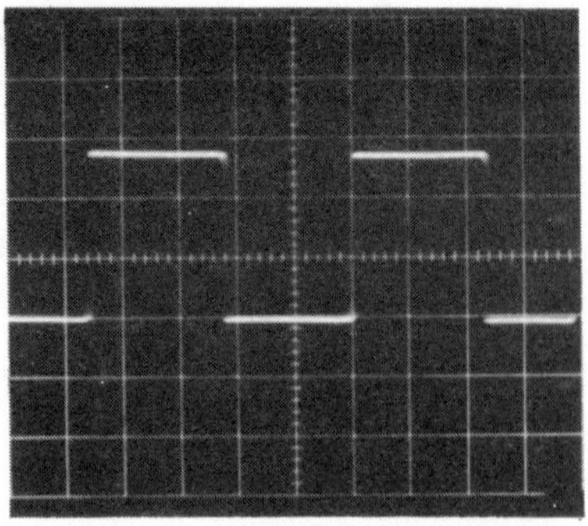

1.1(b)

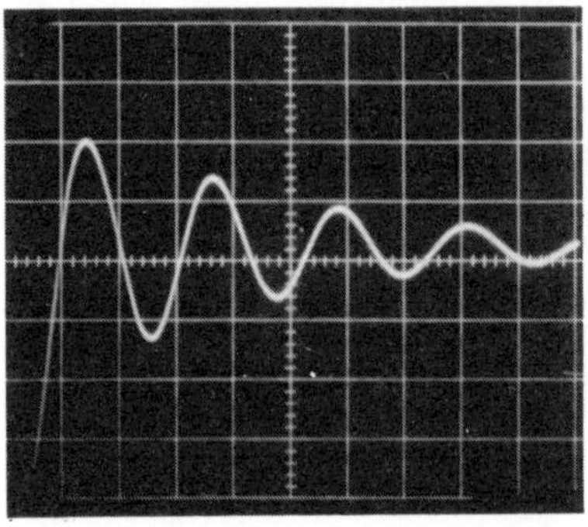

1.1(c)

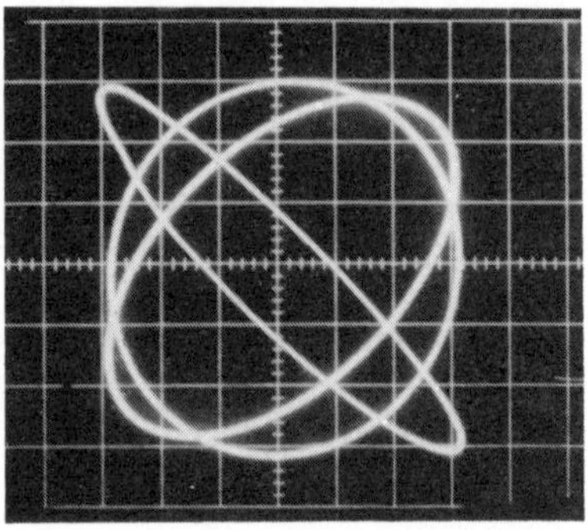

1.1(d)

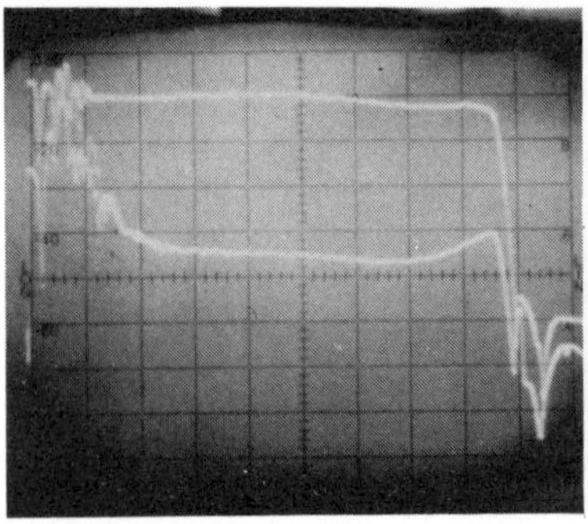

1.1(e)

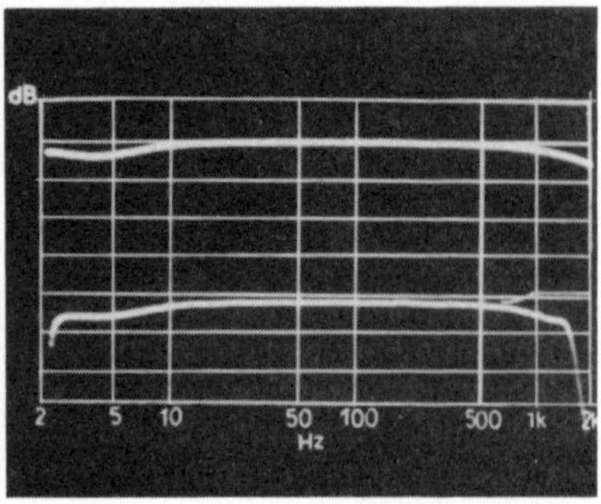

1.1(f)

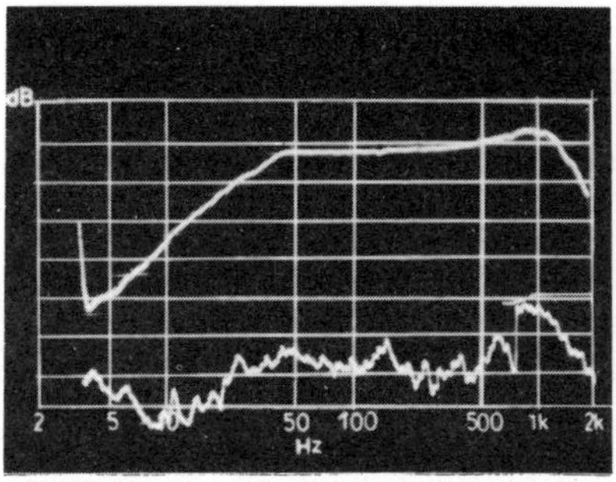

1.1(g)

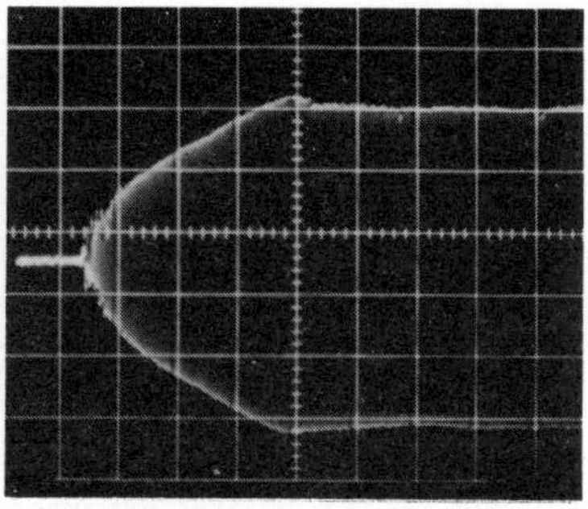

1.1(h)

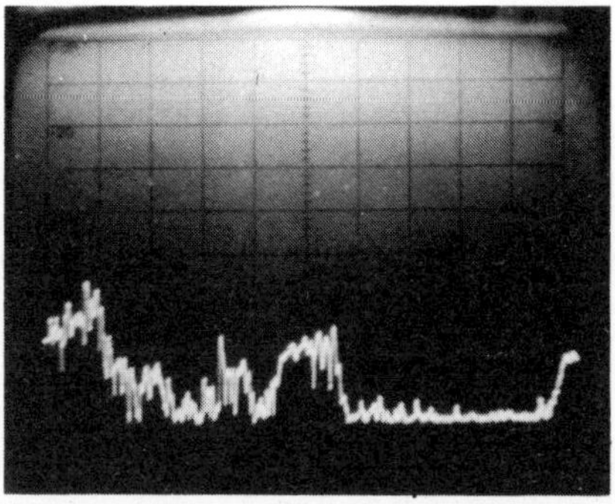

1.1(i)

Fig. 1.1: Examples of some interesting hi-fi equipment parameters. (a) Intermodulation products generated by a hi-fi receiver, scale 2kHz/10dB/div. (b) 1kHz squarewave response of hi-fi amplifier. (c) Ringing produced by a hi-fi amplifier across a loudspeaker load. (d) phase shift parameters of a hi-fi amplifier. (e) Frequency response top and stereo separation below of f.m. tuner, scale 10dB/div. vertically and logarithmic sweep from 20Hz-43kHz horizontally. (f) Frequency response top and stereo separation below of the pickup input of a hi-fi amplifier, scale 5 and 10dB/div. vertically and sweep Hz x 10. (g) Frequency response top and stereo separation below of pickup playing a test tone of constant amplitude to 400Hz and constant velocity to 20kHz, scale 5dB/div. vertically and sweep Hz x 10. (h) Run-up time of turntable unit, scale 0·5S/div. horizontally. (i) Rumble spectrogram of high quality direct-drive turntable unit ref. 10cm/S recorded velocity, scale 10dB/div. vertically and 10Hz/div. horizontally with 1Hz analysing filter.

sound quality is concerned. For example, if a loudspeaker suffers an idiosyncrasy at the treble end, say, and the pickup is similarly harassed, the net result will be an unnecessary emphasis of the treble end affliction. By tutored choice of pickup with respect to the loudspeaker, reproduction of much more palatable quality can be achieved. The scheme is to seek a pickup which has a tendency to cancel rather than enhance the loudspeaker's problem.

Headphone Sets

Headphone sets should also be partly selected by audition rather than by the technical parameters alone. You will, of course, want to be assured that the headphone set will suit your amplifier or receiver. The parameters of the spec will tell you this, but they cannot express how the headphone set will sound to you.

Technical compatibility is high in this area also; pretty well any contemporary headphone set will work adequately by plugging it into the jack socket of almost any contemporary amplifier or receiver. Other aspects are user's comfort and the degree of attenuation afforded by the design to extraneous sounds.

Some designs let through more outside sounds than others. They are designed deliberately to do this; some people find that total intimacy and hence almost complete sound cut-off from the outside world disconcerting when listening from a headphone set.

Tape Machines

Tape machines (both reel-to-reel and cassette), although often used in a hi-fi system, have a greater autonomy than the other items and are often chosen for their own particular facilities and merits from the spec parameters alone. The selection is rather less influenced than the other items by quality compatibility.

Technically, a high degree of compatibility obtains between the latest *decks* (as distinct from complete recorders with their own power amplifiers and loudspeakers) and modern amplifiers and receivers, always provided, of course, that the latter are equipped with suitable inputs and outputs for tape replay and recording.

Record Decks

Record decks, too, are fairly compatible and there is generally less of a problem with regard to cartridge/arm compatibility than there is when the arm and cartridge are purchased separately. This is because some of the technical parameters of the cartridge have to be carefully chosen to suit certain parameters of the arm, and *vice versa.*

FM Aerial

The parameters of the f.m. aerial become critical when the requirements are for fringe area reception and DX-ing, and when there are certain problems associated with the receiving site. For example, in hilly country or where lofty buildings and structures lie in relative proximity to the receiving aerial, the parameter of high directionality may well be desirable to minimise the aerial's response to the signals so reflected.

This allows the aerial to be carefully orientated for maximum discrimination against the reflected signals relative to the wanted, direct signal. The high

Fig. 1.2: Stereo cassette deck by Japanese Akai, Model CS-707D, which includes Dolby noise reduction, memory and bias selection for different tapes.

response of the aerial to the reflected signals, which arrive a small fraction of a second after the direct signal, gives rise to a condition known as *multipath distortion,* which can severely mar the quality of the reproduction and impair the stereo separation.

Where interference from passing cars on a main road can also cause trouble, there is also a good case for a directional aerial. The aerial must always be regarded as an important item of the hi-fi system, for it is economic folly spending a lot of hard-earned pounds on a state-of-art f.m. tuner or receiver and then to feed it with poor or indifferent signals. This is tantamount to running a high performance motor car on low quality petrol!

The parameters of the spec can thus provide a good deal of information on the performance of the various items of the hi-fi setup; but they cannot always tell how well — or not so well — the system will sound and hence audition in the listening room. The parameters are nevertheless extremely important, but are best used in conjunction with overall listening tests.

Parameter Accuracy

The parameters do, though, solely determine the technical excellence of each item, and from them you can prove whether the equipment is behaving as it should (whether it is fully up to scratch). From them you can also find out whether the equipment does really live up to the published specifications.

Nowadays most designers and manufacturers are highly sensitive to ensuring that the parameters are, in fact, accurately stated to avoid confrontation with the Trading Standards people. The parameters are monitored by the hi-fi magazines, the various equipment reviews in this respect doing a great service to the hi-fi buying public.

If you have bought a piece of equipment on the strength of the specification and later discover from your own tests, by those commissioned by an outside authority or by those conducted by a reputable magazine reviewer, that the spec is far from being met, then you certainly have a just cause of complaint to your local Trading Standards Department.

Before you make such a complaint, however, be absolutely sure of your facts and if possible have the parameters in dispute rechecked by a qualified engineer who will issue a signed test sheet. It should be remembered that there are various ways of testing and expressing the various technical parameters, so you must always compare like with like.

Following Chapters

The following chapters in this book endeavour to bring these points into focus. Each chapter is concerned with the specifications and the parameters of specific items of equipment, and each one starts with a typical specification which is analysed parameter by parameter and interpreted in a way that is not specifically technical.

The emphasis is *not* on how to test parameters (these more technical aspects are fully dealt with in my *Audio Technician's Bench Manual,* by the publishers of this book, and some more recent ones in a forthcoming book to be published by Newnes-Butterworth (by Gordon King) entitled *Audio Equipment Tests*) nor particularly how to choose hi-fi equipment, as this subject is fully embraced in my *Choose and Use* books on *Tuners and Amplifiers, Pickups and Loudspeakers,* etc., also by the publishers of this book.

CHAPTER TWO

AMPLIFIER SPECIFICATIONS

THE SPECIFICATION of a hi-fi amplifier is composed of numerous parameters, which may go something like this:

Power Output
25+25W r.m.s. 1kHz 8 ohms; 35+35W r.m.s. 1kHz 4 ohms; 20+20W r.m.s. 40Hz –20kHz; 75W music.

Harmonic Distortion
1% total at rated power; 0·1% at half (–3dB) rated power.

Intermodulation Distortion
0·2% at half rated power (60Hz: 7kHz = 4:1).

Power Bandwidth
20Hz –40kHz (IHF).

Frequency Response
20Hz –30kHz (±1dB).

Damping Factor
30 ref. 8 ohms and 1kHz.

Hum and Noise (volume control zero)
1mV maximum.

Signal-to-noise (S/N) Ratio
Phono 66dB.
Auxiliary 70dB.
Tape 75dB.

Input Sensitivity and Impedances
Phono 2·5mV (47kΩ).
Auxiliary 200mV (100kΩ).
Tape 200mV (100kΩ).

Phono Overload
100mV r.m.s. for 0·5% distortion.

Tape Recording Output
200mV pin (40mV DIN).

Tone Controls
Bass ±10dB at 100Hz.
Treble ±10dB at 10kHz.

Loudness
+8dB at 100Hz and +4dB at 10kHz with volume control set to −30dB.

Filters
Low 30Hz and 12dB/octave.
High 7kHz and 6dB/octave.

Mains Input
240V 50Hz.

The parameters may not be expressed exactly like this and there may be fewer or more of them, depending on the method of testing and on the quality and cost of the amplifier and on the facilities provided. The best way of getting to grips with these parameters is to run through them in turn to see how they can be interpreted. After this the plan will be to investigate the parameters in terms of home requirements.

Power Output

The first on the list, the power output, is a primary parameter of an amplifier. It tells how much urge the amplifier can provide under certain conditions and is akin to the horse power rating of a motor car.

When you buy an electric light bulb or electric fire the main parameter that you particularly look out for is the wattage rating. So it is with a hi-fi amplifier, but in this case it is the power supplied rather than the power dissipated; it is determined as per the simple illustration of power in Chapter 1.

The amplifier is loaded by a large resistor of a suitable power rating (this is called the *output load*), a continuous (sinewave) signal of a specified frequency is coupled to an input of the amplifier and the voltage across the load is measured while it is simultaneously being monitored on an oscilloscope and (possibly) distortion factor meter.

With a stereo amplifier both of the channels are so loaded and driven (the four of a quadraphonic one). This is implied by the example parameter of 25±25W. In this case the loads are 8 ohms. The input *drive* signal is increased until either the peaks of the sinewave are just about to clip or the distortion on the output signal reaches a datum value (usually 0·5% or 1%), which signifies that the *full power* of the amplifier has been reached at that particular frequency. Increasing the drive beyond this point merely increases the clipping and hence the distortion.

The example shows that each channel will yield 25W into an 8-ohm resistive load when the two channels are working together at 1kHz for a distortion of 1% (just at peak clipping). Note that the distortion is indicated by the second parameter quoted and that in this example 25+25W is referred to as the *rated power;* also that this power is available at *1kHz only.*

RMS Voltage

Recalling the simple power calculations in Chapter 1, it will be understood that the voltage across 8Ω for 25W power is equal to the square root of the product of the power and the load resistance (i.e., $V = \sqrt{WR}$), which works out to 14·14V. Thus at the peak clipping point this voltage will be measured across each 8-ohm load.

This is the *root mean square* (r.m.s.) voltage of a sinewave. The 240V of our mains supply is also the r.m.s. voltage. It is the voltage measured on a r.m.s. voltmeter and it is equivalent to the heating power of a d.c. voltage (i.e., from a large battery) of the same value. A sinewave, though, also has a peak value, which is $\sqrt{2}$ *times* the r.m.s. voltage. Thus the *peak* value of 240V mains is 339·41V. From the positive to the negative peak the *peak-to-peak* value is 678·82V!

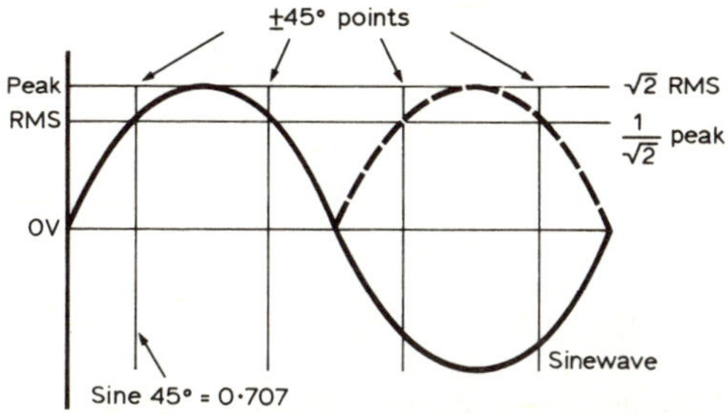

Fig. 2.1: Primary parameters of a sinewave (see text).

Some of the primary parameters of a sinewave are shown in Fig. 2.1. Here we see that the r.m.s. value occurs at four points only when the angles are ±45 degrees and the sine value is 0·707 (i.e., $1/\sqrt{2}$). At any point over a sinewave the *instantaneous* voltage or the *instantanous* current can range from zero to the peak value of the waveform. Likewise the *power.* Into a resistive load, therefore, the power is 0W at the instant when both V and I are zero and *peak* W at the instant when V and I are peak.

Average Power

Using the 14·14V across 8Ω of the exampled 25+25W amplifier, the peak of 14·14V r.m.s. is pretty well 20V. Using this *peak* voltage in the power calculation we get 50W — not 25W. Thus, the power ranges from zero to 50W peak, so the *average* power is 25W, so that when the *r.m.s.* voltage is used in the power calculation the result is the *average power.* This is sometimes erroneously given in the spec as the 'r.m.s. power'. Whenever you read 'r.m.s. power' in a spec. therefore, think *average power!*

If you ever see a spec with a *peak power* parameter halve it to get to the real, *average power*. At one time dodges and artifices such as this were adopted by the ad people to give the false impression of the amplifier being more powerful than a competitor's of possibly lesser, real *average power*. Happily, the Trade Descriptions Act of 1968 has put a penalty on such sharp practices! But still be on the lookout yourselves.

Load Values

Contemporary transistor amplifiers are *constant voltage* devices. This means that the voltage across the load holds fairly constant over a range of load values right down to the lowest value load that the amplifier is designed to accommodate. This results from the use of heavy negative feedback and the resulting low source resistance (see under 'Damping Factor').

The spec may well indicate a range of load values, corresponding to the impedances of latter-day loudspeakers (for the amplifier is designed ultimately to drive a loudspeaker — not to heat resistors or light electric light bulbs!). The example parameter indicates a maximum output of 14·14V r.m.s. across 8Ω (giving 25W average power for each channel). If it is assumed that this voltage holds across 4Ω, then, of course, the power is double or 50W. If the load is 16Ω, then the power is halved (12·5W).

Current Limiting

In spite of the low source resistance of the amplifier, the voltage to peak clipping does tend to fall slightly as the load is reduced, which is why the example parameter shows 35+35W into 4-ohm loads, instead of the expected 50+50W. Nevertheless, the power into the load does rise as the load value is decreased.

The current through the load also rises, of course, as its value is reduced, and as excessive current can blow the output transistors, current limiting is often included in the design, which is another reason why the power does not always double when the load resistance is halved. If there was no such limiting (or fusing) the current would rise infinitely if a short circuit happened to occur across the loudspeaker or wiring under full drive.

Thus, when comparing amplifiers for power yield *always make sure that you are comparing like with like.* That is, that the comparison is made at a common load value — 4Ω or 8Ω, whichever is the most convenient.

Frequency Range

So far we have investigated the power parameter at 1kHz only. These days amplifiers are expected to deliver their *rated power* over a frequency range from 20Hz – 20kHz, or thereabouts. Not all amplifiers are capable of delivering the power available at 1kHz over the whole frequency range.

In the example parameter it will be seen that over 40Hz – 20kHz the power is down to 20+20W. It would undoubtedly be even less than this at 20Hz. In this case the spec writer has deliberately restricted the low-frequency to avoid having to derate the power too much (i.e., over *20Hz* – 20kHz it may be 12+12W!). It will also be noticed that a load value is not given for this part of the parameter. *Always regard with great suspicion any parameter which is incomplete.*

Thus when comparing amplifier powers *always make sure that you are comparing the powers at a common frequency or over a common frequency range.* For example, a parameter going something like 25+25W 8Ω 20Hz – 20kHz is better than one going 25+25W 8Ω 40Hz – 20kHz and far better than one going 25+25W 4Ω 40Hz – 20kHz.

Music Power

The term 'music power' is sometimes included in the power parameter. It is not very meaningful and is less used nowadays. It shows how much relative power the amplifier can deliver on music type signal as distinct from continuous sinewave signal.

The measurement is made using 'chopped' signal, and because the output is not then continuous (the amplifier having a little time to recover between 'bursts') the effective power capability is often increased over the average power, but this depends on the design and adjustment of the power amplifier, including the size of heat sinks used for the output transistors, and on the nature of the power supply.

Moreover, the music rating is commonly expressed as that power of the two stereo or four quadraphonic channels added together. For example, if the average power of an amplifier is given as, say, 25+25W, then the music power would be 50W *plus* the extra which is available due to the use of a non-sinewave signal, which could add another 20% *or more* to the average power rating.

It should always be remembered that the average power of an amplifier is the same sort of power that lights our bulbs and runs our electric fires. It differs in frequency, of course; in the UK the mains frequency is 50Hz (60Hz in American countries), which is at the low-frequency end of the audio spectrum — near G2 on the piano (the eleventh key up from the bass end). Our amplifiers are expected to supply power from the low-frequency of 20Hz or so up to, at least, 20kHz; that is, a range of three decades or ten octaves, at least!

While mains power into a bulb or electric fire is continuous at the particular wattage rating, amplifier power fluctuates widely on music signal. By feeding an amplifier channel with continuous sinewave signal and connecting a bulb in place of the loudspeaker, the bulb would glow continuously and brightly when its voltage and wattage ratings correspond to those of the amplifier.

Feeding the amplifier channel with music signal, from a gramophone pickup or radio tuner, for example, the bulb illumination would pulsate in sympathy with the music. Our amplifiers, of course, are not designed for this. Their job is to supply current to the connected loudspeakers.

Amplifier/Loudspeaker Interface

While loudspeakers *are* current operated, their sensitivity in terms of *sound pressure* output at a given distance and under stipulated conditions is referred to an imput *voltage* of shaped noise signal which has some of the features of music signal. The actual *power* abstracted by a loudspeaker partly depends on its modulus of impedance (Z). Chapter 6 gives information on this, but some aspects of the amplifier/loudspeaker interface will be best looked at now.

It has already been noted that a hi-fi, negative feedback amplifier represents a constant voltage source. Thus, at a given output voltage the current flowing through a connected loudspeaker depends on Z. Impedance Z means that the load includes reactive components (i.e., capacitance and inductance).

The value of Z is therefore not constant over the frequency range, as pure resistance (R) is. It is usually the value measured at 400Hz or 1kHz — called the *nominal impedance*. At the low-frequency resonance of the loudspeaker Z can rise a number of times above the nominal impedance — called the *motional impedance*. There may also be other, less dramatic peaks and troughs at higher frequencies, depending on the overall system design and the frequency-dividing networks used.

The current, and hence the power, thus falls when Z rises and rises when Z falls. For this reason alone it is virtually impossible to calculate the average power taken by the loudspeaker on wide-band signal.

Power Factor

Another reason why the power delivered to a loudspeaker is not as would be expected from the impedance is because owing to the reactive components of the load the current through it and the voltage across it do not rise and fall precisely in step at all frequencies. We have seen (Chapter 1) that with a pure R load the power is W = VI. With a Z load the power is W = VIcosø, where cosø is the cosine of the angle of lead or lag of the current with respect to the voltage, which is the *power factor*.

At some frequencies the ø of certain loudspeakers can swing to 40-60 degrees. Assuming a swing of 60 degrees, the cosine of this is 0·5, which means that the power is half that which would be produced into a pure R load, when the Z and R values are the same!

Moreover, if the ø is large when the Z is low, some amplifiers are 'triggered' into current limiting before the maximum voltage that can be obtained across a pure R load is reached, which, of course, puts a limit on the maximum sound pressure output relative to a given level of distortion.

For these reasons, therefore, it would be far better to have our amplifiers rated in output voltage across specified loads rather than in output power, which merely signifies the potency of the amplifier in heating pure resistance!

Distortion

This parameter was introduced in the foregoing section. When the rated output of an amplifier is reached the distortion is round 0·5 to 1% corresponding to the threshold of peak clipping. Any attempt to increase the output would result in a swift and considerable rise in distortion.

Fig. 2.2 oscillograms depict sinewave signal on one trace at a level causing severe clipping of one peak and music signal at three levels on the other trace. Low level music well below clipping is shown at (a). Higher level music but still below clipping is shown at (b).

Much higher level music this time taking the amplifier into peak clipping is shown at (c). While (a) and (b) would give a palatable sound, (d) would be very rough and disconcerting. Below the rated output (a) and (b) the distortion should be very small. The example spec shows a distortion of only 0·1% at half (−3dB) the rated output.

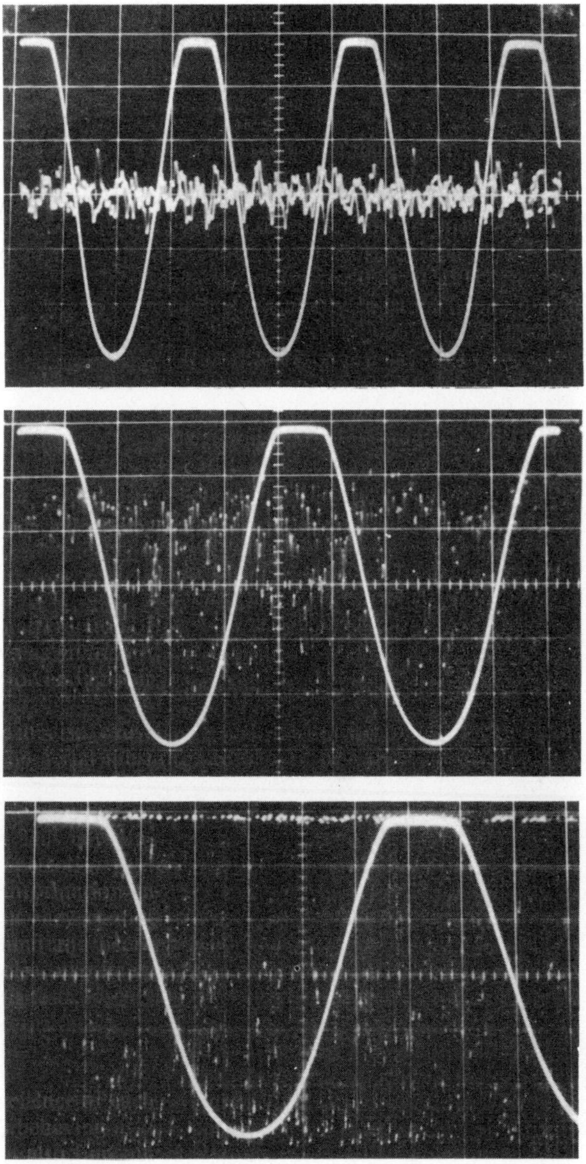

(a)

(b)

(c)

Fig. 2.2: Oscillograms showing peak-clipped sinewave on one trace and music signal on the other trace. (a) Low level music signal well below clipping. (b) Higher level music signal but still below clipping. (c) Very high level music signal running into peak clipping and hence severely distorted.

Harmonic Distortion

This is harmonic distortion which is measured by applying a very pure (distortion free) sinewave input at the required frequency and turning up the output for the required *voltage* across the load. A special instrument (distortion factor meter) is then used to delete the fundamental so that only the sum of all the harmonics and noise components remains. This is then compared in terms of a *voltage* ratio or percentage with the original output *voltage.*

For example 0·1% means that the summed voltage of the harmonics and noise components is 1,000 times less than the voltage of the fundamental. If the fundamental corresponds to 20V, then the harmonics, etc. would correspond to a mere 20mV (i.e., 0·02V). There are other ways of measuring harmonic distortion with a wave or spectrum analyser (see *Audio Technician's Bench Manual*).

Some of the latest top-flight amplifiers boast an even lower distortion, but the percentage distortion is often related to frequency, usually being less at 1kHz than at low and high frequencies at the spectrum extremes.

A clipping amplifier produces a whole train of harmonics. Based on a 1kHz fundamental, we get the second at 2kHz, the third at 3kHz, the fourth at 4kHz, etc. Some of the higher order, odd harmonics are much more disconcerting than higher *level* low order, even harmonics. A fair measure of second and possibly third harmonic distortion can be tolerated, but dissonant harmonics referred to middle C and taken as 250Hz for convenience include the seventh at 1,750Hz, the ninth at 2,250Hz, the eleventh at 2,750Hz and other odd ones up to about the twenty-fifth at 6,250Hz. All these are less liked by the ear even when their amplitudes are less than those of lower order, even harmonics.

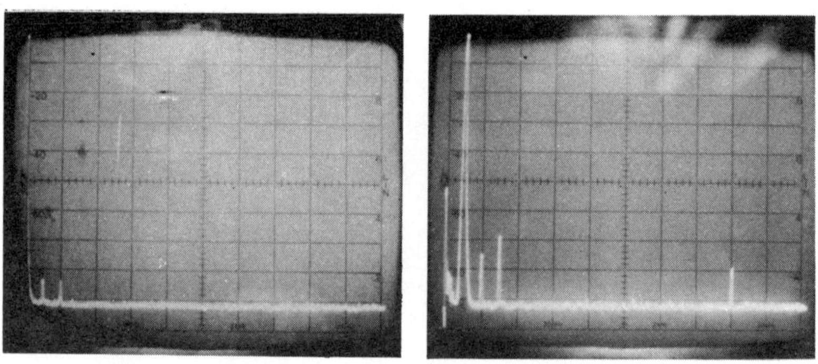

(a) (b)

Fig. 2.3: Harmonic spectrograms ref. 1kHz OdB driving signal. (a) Very good result showing second and third harmonics at −83dB (0·007%). (b) Less dramatic though good result showing second harmonic at -75dB (0·017%) and third harmonic at -96dB (0·035%), with spurious signal of -80dB (0·01%) at 16kHz. Scale 10dB/2kHz/div.

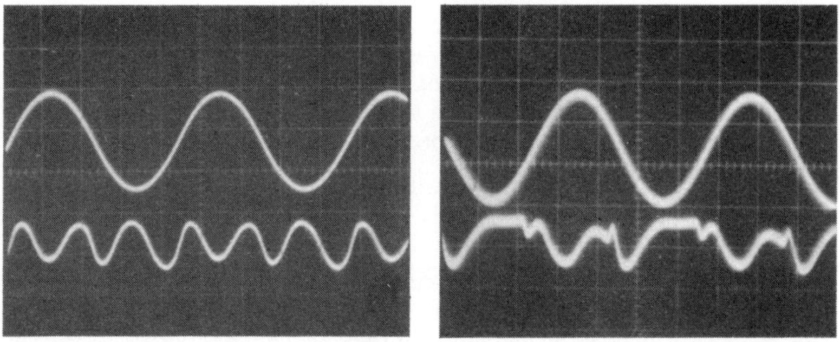

(a) (b)

Fig. 2.4: Distortion factor oscillograms with the driving sinewave on one trace and the residual on the other trace. (a) Mostly second and third harmonic. (b) Showing crossover distortion and the resulting high order harmonic (see text for evaluation).

Spectral Analysis

A well designed amplifier operating below clipping produces mostly second and third harmonics, and a couple of examples are given in Fig. 2.3, (a) showing the remarkable result from the Pioneer SA9900 amplifier, where the second and third harmonics are each at the astonishingly low level of —83dB (0·007%), and (b) a less dramatic but nevertheless good result of —75dB (0·017%) second and —69dB (0·035%) third, with a —80dB (0·01%) spurious signal at 16kHz.

Only this sort of spectral analysis can identify the individual harmonics, spurious signals and their amplitudes with respect to the driving signal, which is 1kHz in the examples set to 0dB datum.

The sum of the harmonics and noise from the output of a distortion factor meter is shown merely as a waveform on an oscilloscope, but the nature of the waveform sometimes gives a clue as to the order of the harmonics. Examples are given in Fig. 2.4, (a) where the harmonics are low order (mainly second and third), and (b) where the 'peaky' parts of the distortion waveform indicate higher order components.

Display (b), in fact, indicates the presence of so-called *crossover distortion* which occurs in some transistor amplifiers at low output owing to incorrect biasing or poor design of the power amplifiers. This distortion, being of high order, is dissonant. At low volume, therefore, the amplifier responsible for (b) would possibly audition less favourably than the amplifier responsible for (a), even though the distortion factor *percentage* might be far greater at (a) than at (b).

This kind of detailed distortion information is rarely given in manufacturers' specifications, but it can be found in the comprehensive reviews published by reputable hi-fi magazines.

It will be understood, of course, that the distortion of all the stages in the amplifier is measured when the test signal is applied at the input and measured at the output. In some specs the distortion may be given separately for the preamplifier and power amplifier sections.

You may sometimes see the term *total harmonic distortion* or its abbreviation *THD*. This is virtually the same as distortion factor since it means the total or sum of all the harmonics expressed as a percentage of the voltage of the fundamental signal.

Intermodulation Distortion

The next parameter listed is *intermodulation distortion*. Both harmonic and intermodulation distortion (IMD) are caused by amplitude non-linearity. This merely means that the amplitude of the signal voltage across the output load fails precisely to track a change in amplitude of the input signal voltage.

For absolute tracking, the input/output characteristic would need to be a perfectly straight line as shown at (a) in Fig. 2.5. In practice the characteristic is slightly curved, as shown exaggerated at (b). Thus, (a) signifies the characteristic of a distortion-free amplifier (an impossibility) and (b) the characteristic of an amplifier producing high values of THD and IMD.

A hi-fi amplifier of very low distortion (Fig. 2.3) has a characteristic which is very, very slightly curved within its dynamic range (minimum to maximum usable output). However, any amplifier, no matter how high the fidelity, has a characteristic which curves violently at the top end of the dynamic range (high output). This is caused by the overload condition, where the output signal is clipped, as shown at (c) in Fig. 2.2.

The amplitudes and the orders of the harmonics are related to the nature of the curve. The amplitude non-linearity also gives rise to sum and difference frequencies when two or more frequencies are passing through the amplifier

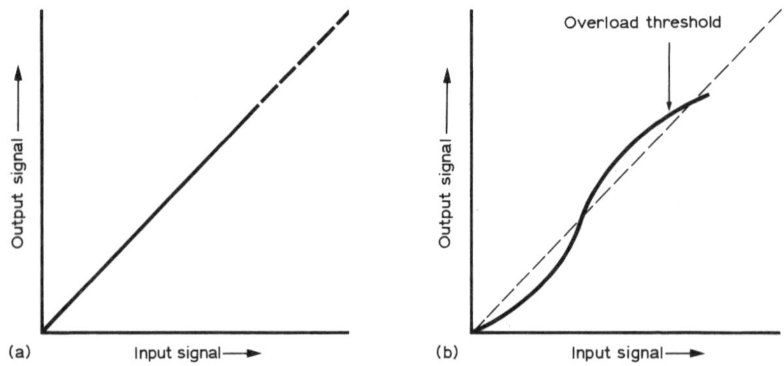

Fig. 2.5: Input/output characteristics. (a) Perfect straight line amplitude linearity and (b) non-linearity referred to the straight line condition (shown by broken-line characteristic). All amplifiers are subject to a degree of non-linearity which causes harmonic and intermodulation distortion, but the curvature within the dynamic range of a hi-fi amplifier is very small. Hence the low distortion yield of such amplifiers. At overload all amplifiers show severe non-linearity.

together. Examples from two signals of f1 = 60Hz and f2 = 7,000Hz are f1+f2 = 7,060Hz, f2−f1 = 6,940Hz, f2−2f1 = 6,880Hz, 2f2−2f1 = 13,880Hz, f2−3f1 = 6,820Hz, etc. One can well imagine the incredible number of *intermodulation products,* as they are called, which could arise from more than two frequencies (music signal, for example) passing through a non-linear amplifier!

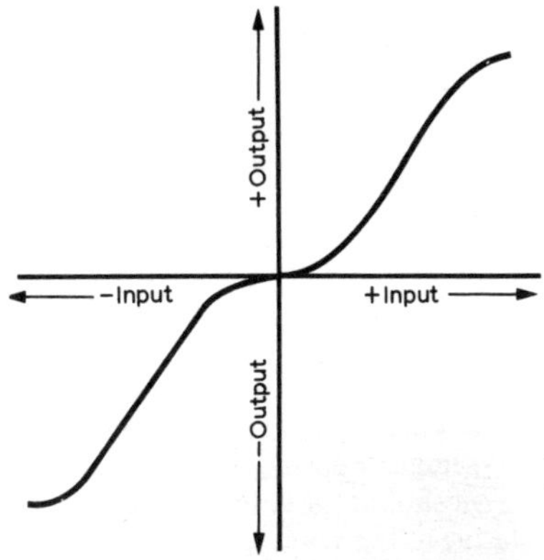

Fig. 2.6: This sort of crossover non-linearity is responsible for crossover distortion in some quasi-Class-B amplifiers (see text).

Crossover Distortion

The difference between simple harmonics and intermodulation products is that while the former are harmonious (musical instruments themselves produce harmonics which are responsible for the timbre of the sound) many of the latter are highly dissonant (having no acceptable musical relationship).

However, a well designed amplifier of low THD should not generate much more IMD within its dynamic range; but like THD, the IMD will rise swiftly at the overload threshold. With some of the less detailed quasi-Class B amplifiers there is also the tendency for the distortion to rise at the low end of the dynamic range. This is caused by mild discontinuity of the input/output characteristics of the two push-pull output transistors at the crossover point, as shown exaggerated in Fig. 2.6. This reveals that at low output levels significant non-linearity occurs, also see (b) in Fig. 2.4.

The resulting *crossover distortion* is exposed more at low output by using an IMD test than a THD test, but not all specs give an IMD parameter at very low level. In the example spec it will be seen that the parameter relates down only to *half* rated power. Good test reports and reviews are sometimes more critical in this respect, so to get the full story on distortion performance they are well worth investigating.

IMD Spectrograms

The spectrograms in Fig. 2.7 show IMD products of two amplifiers from equal amplitude driving signals of 15 and 16kHz. No product is much greater

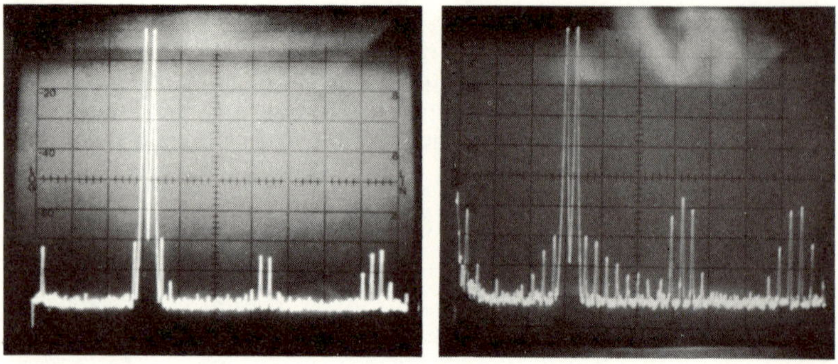

(a) (b)

Fig. 2.7: Spectrograms showing IMD using twin tones (i.e., two driving signals of equal amplitude), (a) low distortion amplifier and (b) amplifier producing high distortion. Scale 5kHz/10dB/div.

than —70dB (0·03%) at (a), but at (b) there is a product up to —56dB (0·16%). Displays such as these given in test reports and reviews will indicate the voltage of each driving signal and the type of output load employed.

You should note in particular the amplitude and number of IMD products relative to the voltage of the driving signals and the type of load used, remembering that a loudspeaker type load can produce more distortion than a resistive load, especially at the top of the dynamic range where the phase angle of the load may be resulting in premature limiting of the amplifier, as already explained.

The spectrogram of a badly distorting amplifier with driving signals at about 100Hz and 1·4kHz is shown in Fig. 2.8. Notice here the large number of products and the large amplitudes of some of them.

In general, then, look for an amplifier with the lowest values of THD and IMD, but remember that an amplifier of relatively large THD may sound

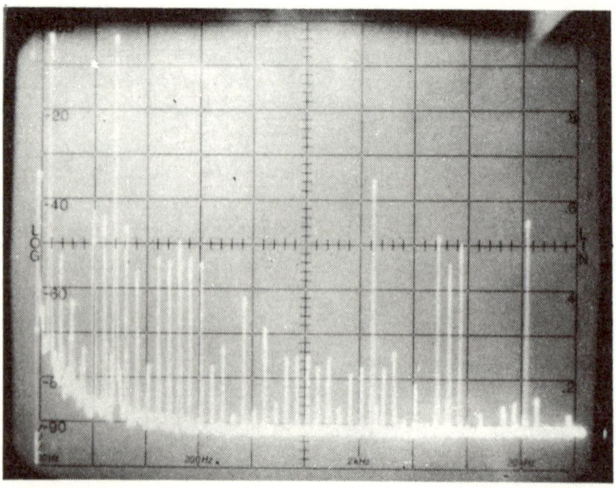

Fig. 2.8: Very bad IMD. Scale 2kHz/10dB/div.

better than one of smaller THD when the harmonics of the latter are dissonant. A better correlation between the measured results and the 'sounding' of an amplifier is given by IMD, particularly when expressed by a spectrogram (also see Chapter 9).

Power Bandwidth

This refers to the frequency range over which a given output *power* can be obtained with respect to a specified level of distortion. A resistive load is employed and the distortion datum is usually 0·5 or 1%. The more recent way is to refer the bandwidth to the *rated power* of the amplifier at 1kHz. Thus, if the rated power at 1kHz is, say, 20+20W, then the power bandwidth refers to those frequencies below and above 1kHz where 20+20W can be obtained at a distortion no greater than, say, 1%, as shown in Fig. 2.9.

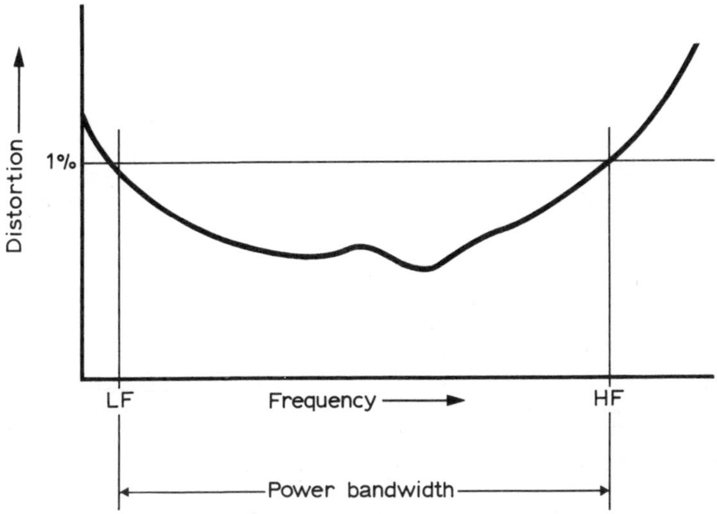

Fig. 2.9: Definition of power bandwidth. In this example the bandwidth is from the low-frequency (LF) to high-frequency (HF) points where the distortion is 1% at the amplifier's rated power.

The *half power bandwidth* refers to the low- and high-frequency points where the distortion is no greater than 0·5 or 1%, but this time at *half* the rated power of the amplifier. The bandwidth is greater from a given amplifier under these conditions than when the bandwidth is referred to the rated power.

Although the 'energy' of music signal is less at the low and high frequencies than at mid-spectrum (music signal tends towards maximum at 700Hz), it is desirable for the amplifier to be capable of delivering its full rated power without clipping and hence distorting down to at least 40Hz to at least 10kHz.

It is noteworthy that clipping distortion from amplifiers unable to deliver the rated power at low and high frequencies (including low power amplifiers) is a common cause of thermal damage to loudspeaker drive units.

Frequency Response

Whereas the power bandwidth refers to the frequency range at power, the frequency response is a lower level parameter, it indicating how the output varies over the frequency range at 1W or less output.

A hi-fi amplifier should boast a virtually flat frequency response from at least 20Hz bass to 20kHz treble. The response is usually plotted in the form of a curve relative to 0dB at 1kHz, as shown in Fig. 2.10. Here the output remains essentially constant from 20Hz to in excess of 20kHz; at 10Hz it is 5dB down and at 100,000Hz 10dB down!

If 0dB corresponds, say, to 1V across the output load, then at points where the response is −5dB the *voltage* will be 0·562V and at −10dB it will be 0·316V. Based on *power* where 0dB corresponds to 1W, say, at −5dB the power will be 0·316W and at −10dB it will be 0·1W. Half power corresponds to −3dB, as already noted elsewhere.

The decibel (dB) scale is thus a logarithmic scale of voltage or power *ratios.* The dB scale is used in hi-fi because it correlates much more than a linear scale to the way that we hear. More details about the way that the dB is used in hi-fi can be obtained from the companion volumes *Tuners and Amplifiers, Pickups and Loudspeakers, Improving Your Hi-Fi, ABC of Hi-Fi, Audio Technician's Bench Manual,* etc.

Half Power Point

The −3dB, half power point is significant in frequency responses, for some responses may be expressed as, for example, 10Hz−40kHz (−3dB). This means that the frequency range lies between the points on the response where

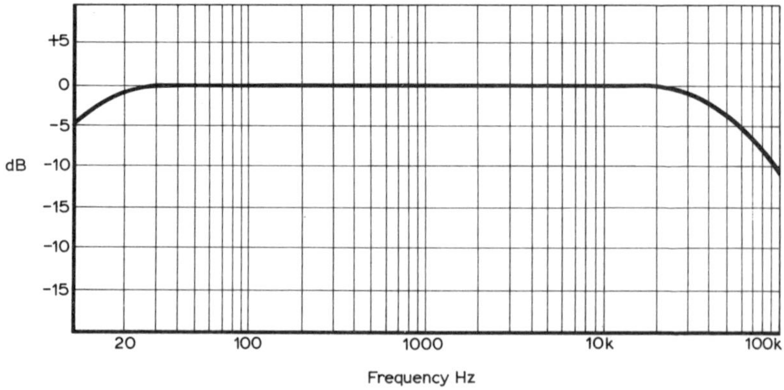

Fig. 2.10: Example frequency response which can be defined as 15Hz to 45kHz −3dB (see text).

the output has fallen by 3dB ref. 0dB at 1kHz. In Fig. 2.10 this would be between about 15Hz and about 45kHz. A plotted curve is better, though, since it shows unambiguously how the output rises and/or falls between the two −3dB points.

The frequency response is modified by the filters, loudness control, tone controls, etc. The accuracy of the RIAA equalisation for the gramophone pickup is also shown by a frequency response. For this measurement an RIAA recording filter is connected in series with the test signal source.

Filters

Although the intrinsic frequency response should desirably be 'flat' over the spectrum with minimal undulations, there are times when better reproduction can be obtained by deliberately reducing the output at certain frequencies. One example of this is at the high-frequency end of the spectrum when reproducing programme material which is rather 'noisy' — for example, from a scratched gramophone record or 'hissy' f.m. stereo radio programme.

By rolling-off (as it is called) the hi-frequency end of the spectrum the clicks from the record and the 'fizz' from the radio tuner can be reduced. This is because such disturbances contain components which extend well beyond the upper musical frequencies. However, if the roll-off is started too early the treble of the music will be diluted and the reproduction will lack the sparkle and attack associated with hi-fi.

When the unwanted noises are bad, a compromise between music quality and roll-off frequency has to be adopted, which is why some amplifiers allow you to switch the roll-off frequency. The frequency response in Fig. 2.11 shows two high-frequency roll-offs, one starting round 9kHz and the other round 20kHz. The latter affects the music quality far less than the former, but the former will reduce background hiss more than the latter.

The turnover frequency is that frequency where the response is 3dB below the response at 1kHz and the steepness of the roll-off slope is given as so many *decibels per octave*. The least slope is 6dB/octave, and this can catch some of the higher wanted music frequencies. The greater slope of 12dB/octave is generally preferable for filtering, for the turnover frequency can then be made higher with the swift fall blocking most of the noise components. Higher rates are sometimes used, but filters of this kind tend to impair the transients of the music, causing 'ringing' and overshoot with a consequent tonal distortion on the reproduction.

In addition to turnover frequency switching, some amplifiers are also equipped with a control for varying the roll-off rate. Such filters, cutting the high frequencies, are known technically as low-pass filters because they pass the lower frequencies below the turnover frequency.

The filter at the other end of the spectrum is called a low-frequency, bass or high-pass filter. Except for large organs, there is not much in the way of music content below about 20Hz, so it can sometimes be undesirable for the amplifier to respond much below this frequency. Warps on gramophone records, for example, can produce sub-bass frequencies which merely cause the cone of the bass loudspeaker unit to pump in and out, thereby using up amplifier power unnecessarily and contributing to the distortion.

A bass filter round 20Hz turnover avoids these problems, while also reducing the lower frequency rumble of turntables. A higher turnover is needed, though, to eliminate all the rumble, particularly of a poor deck, and the roll-off rate must be greater than the minimal 6dB/octave to prevent the wanted low bass music frequencies from being unduly attenuated. Responses of such filters are given in Fig. 2.12.

The parameters of filters may thus either be expressed by frequency/amplitude plots or by figures, such as 8kHz−3dB and 12dB/octave, 12kHz−3dB and 6dB/octave, 30Hz−3dB and 12dB/octave, etc.

Tone Controls

A tone control is really a filter of 6dB/octave maximum roll-off rate which is adjustable to give both cut and boost. It is also sometimes possible to switch the turnover frequency for a greater overall versatility of control. All hi-fi amplifiers have at least two tone controls, one for bass and the other for treble. Some have an additional control giving lift and cut round 1 or 2kHz.

Parameters may be expressed by frequency/amplitude plots as shown in Fig. 2.13, at (a) for unswitchable controls and (b) for two pairs of turnover frequencies, or by figures. Typical parameters for bass and treble are bass ±10dB at 100Hz and treble ±10dB at 10kHz. In Fig. 2.13 the middle sweep is with both controls at 'neutral' (flat), the top sweep with bass and treble and full lift and the lower sweep with bass and treble at full cut.

It can sometimes help to apply a little boost or perhaps cut at the bass or treble to improve the reproduction in relation to the programme information or the acoustics of the listening room, including the loudspeakers; but in general excessive boost or cut is unnecessary.

Tone controls tend to provide an undesirably wide range of control. Tone controls can — and do — introduce distortion of various kinds, and because of this certain hi-fi freaks are tending to prefer amplifiers without such controls or, at least, with a switch to bypass the tone control stages.

Loudness

When the loudness switch is operated, the frequency response of the amplifier changes as the volume control is turned down to reduce the sound intensity. What happens is that the middle of the spectrum becomes more attenuated than the bass or treble ends as the control is retarded, with the result that the bass frequencies, and to a smaller degree the treble ones, are boosted relative to the middle frequencies. This tends (unnaturally!) to compensate for the way that we hear. For instance, as the sound intensity is reduced so the 'sensitivity' of our hearing falls at the bass and to a smaller extent at the treble.

Our sample spec shows that when the volume control is turned down from maximum so that the gain is reduced by 30dB (a thousand times power reduction) a boost of 8dB occurs at 100Hz and a boost of 4dB at 10kHz. The effect is revealed graphically in Fig. 2.14. The boost increases as the volume

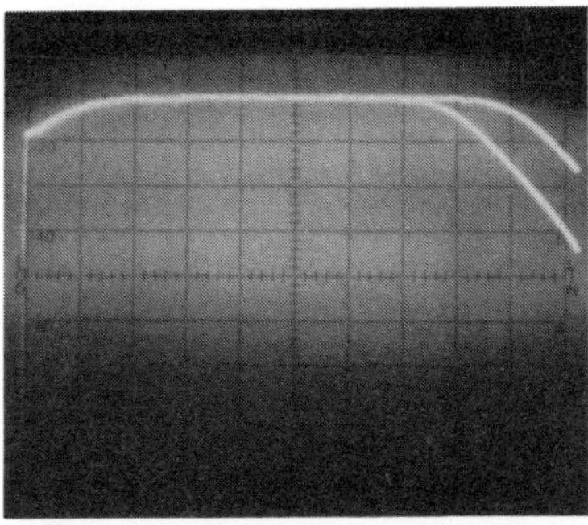

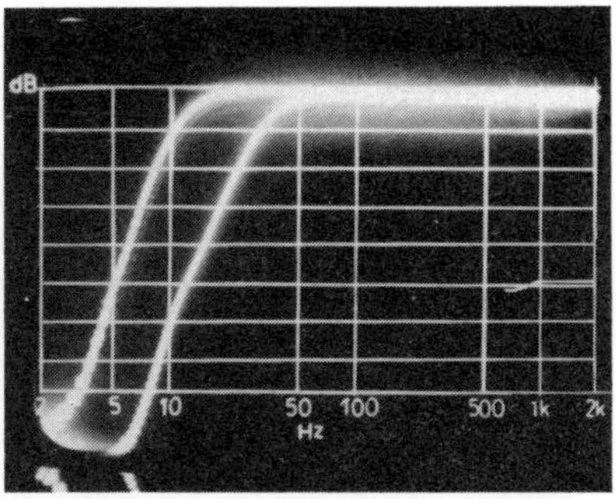

Fig. 2.11: (*left*) Frequency response showing two high-frequency roll-offs and a smaller low-frequency roll-off. The high-frequency roll-offs correspond to two, switchable low-pass filters. Scale 10dB/div. vertically and three divisions per frequency decade horizontally, sweep logarithmic 20Hz – 43kHz.

Fig. 2.12: (*right*) Bass filter (i.e., high-pass filter) response at two turnover frequencies. Scale 5dB/div. vertically and Hz x 10 logarithmic sweep 20Hz – 20kHz.

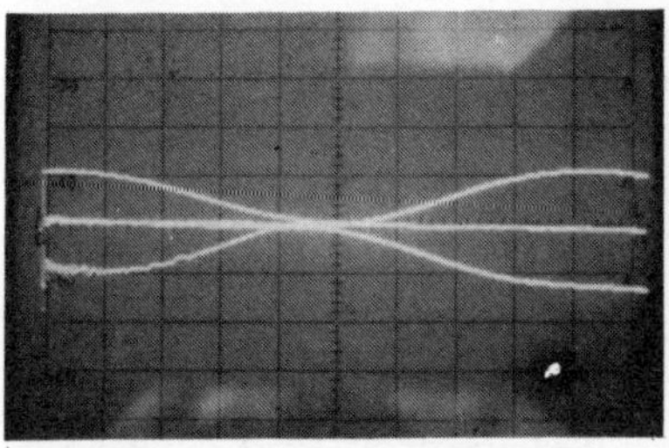

(a)

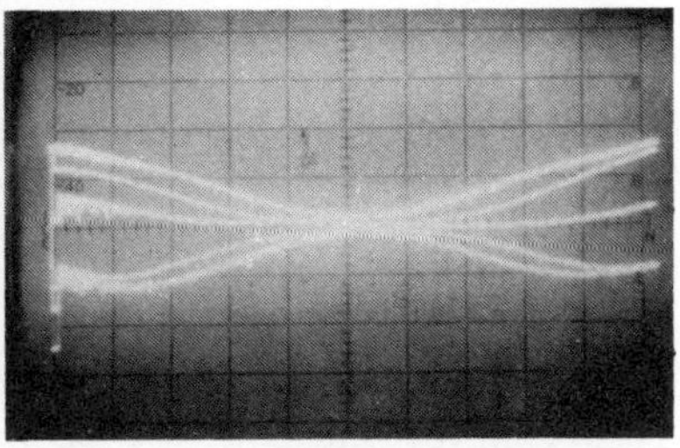

(b)

Fig. 2.13: Tone control responses, (a) fixed turnover frequencies and (b) switched turnover frequencies, scale as Fig. 2.11.

control is further turned down and decreases as it is advanced — but, normally, only when the loudness switch is on. With the switch off the volume control works normally without frequency compensation.

Some amplifiers and receivers, though, notably European (not British) makes, are equipped with unswitchable loudness permanently 'ganged' to the volume control. This is not a particularly desirable attribute, but it is possible to counteract the effect by turning back (for cut) the bass and treble controls when listening at low level.

Damping Factor

When a loudspeaker receives from the amplifier a sudden pulse (e.g., music transient) the cone is swiftly deflected. After the transient has passed, the cone will return to its steady-state position. However, if the loudspeaker is inadequately 'damped' the cone will overshoot and move to and fro before finally coming to rest. This tends to 'colour' the reproduction, sometimes emphasising 'bass boom' or 'one-note bass'.

When the cone is vibrating due to lack of damping, the loudspeaker acts rather like a dynamo. Electricity flows out of the loudspeaker into the amplifier (an effect which can cause some amplifiers to produce interface intermodulation distortion — IID). If the output terminals of the amplifier 'look' like a very low resistance then the cone vibrations are quickly suppressed or damped. An amplifier with a high damping factor is endowed with a very low resistance output (just the matter of a few milliohms) and hence the loudspeaker is nicely damped.

The damping factor is numerally equal to the loudspeaker's impedance divided by the amplifier's output impedance. Thus with an 8-ohm loudspeaker, an amplifier with a source impedance of 0·1Ω (100 milliohms) will have a damping factor of 80.

It is important for the damping factor to remain reasonably high at very low frequencies for it is at the bass end when high damping is necessary to reduce the so-called 'one-note bass' effect of certain loudspeakers. The actual damping, of course, is always less than that provided by the amplifier because in series with this damping will appear the resistance of the loudspeaker's speech coil, frequency-divider components and loudspeaker connecting cables. It appears that there is little point in providing an amplifier damping factor greater than about 30 owing to these series effects.

S/N Ratio

While the maximum output of an amplifier determines the top end of the dynamic range, the lower end depends on how much noise the amplifier is producing. If the noise is relatively high, then very quiet parts of the music will be masked and there will be a loss of information. The noise is measured, often through a 'weighting' filter which simulates the way that we actually discern the noise (see Chapter 3), relative to the amplifier's full output. It thus gives an expression of *dynamic range*.

Ratios of 66dB phono (magnetic pickup), 70dB auxiliary and 75dB tape are perfectly acceptable. The measurements are made with the appropriate input loaded and with the volume control at maximum. Under normal music conditions better ratios generally occur because we rarely operate with the volume control flat out!

Residual Hum and Noise

This measurement is similar to S/N ratio but it is made with the volume control turned right down so that it is the hum and noise of the stages after the volume control which in this case are evaluated. The actual hum and noise voltage across the amplifier's output load is commonly measured which, in the sample spec, is 1mV. This corresponds to the mini-power of 1·25 x 10^7W, so you would have to get very close to the loudspeaker to hear the result!

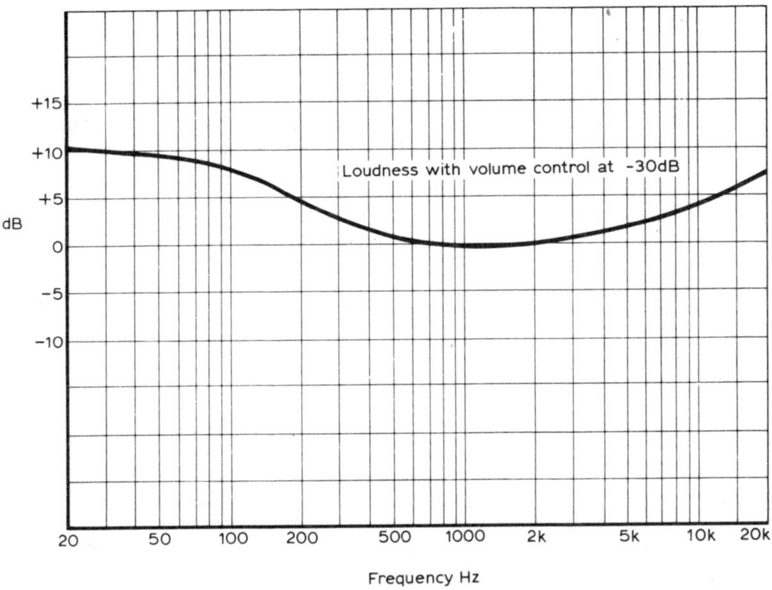

Fig. 2.14: Loudness response with volume control set at −30dB.

Input Sensitivity and Impedance

This parameter tells you the level of signal at mid-spectrum (1kHz) that the amplifier needs at any of its inputs to give its full output with the volume control at maximum (all other controls at 'neutral' or switched out). It also tells you the impedance or resistance that the inputs present to the programme sources.

The particular programme source should give an 'average' signal fairly close to the appropriate input sensitivity and the impedance or resistance reflected from the input should not adversely affect the signal from the source. In the latter respect, it is best for the impedance or resistance of the input to be greater than that of the source. The frequency response of the source is then affected least. The other way round, the capacitances of the screened signal-coupling cables can cause high-frequency roll-off of the source signals.

The absolute performance of a magnetic pickup (cartridge) is often determined by the 'load' it 'sees' from the phono (pickup) input. The average cartridge load is round 47kΩ, which is why the pickup input impedance is close to this value. However, this is less applicable to low impedance cartridges, such as moving-coil types. On the other hand, certain relatively high impedance cartridges are very critical to load, including shunt resistance *and* capacitance, and some of the better class amplifiers and hi-fi receivers cater for this by input impedance switching.

In general, the input sensitivities and impedances of latter-day amplifiers and hi-fi receivers are carefully tailored to suit current signal sources so there should not be too much trouble in securing a fair coupling match, and they are switchable on some amplifiers.

Pickup Overload

Although the 'average' output of a pickup may only be some 5mV, to give the full dynamic range of the record the signal level will vary over some 60dB relative to this 'average'. At very low outputs the overall noise becomes significant, while at very high outputs the distortion and overload take effect. Thus to avoid serious distortion at high outputs the RIAA preamplifier must be capable of handling the signal without overloading. This is expressed at 1kHz by the pickup overload parameter.

In the sample spec it will be seen that although the input sensitivity at pickup is 2·5mV, the input will accommodate up to 100mV before serious overload becomes apparent. These are r.m.s. values. Peak values (that is the r.m.s. value *times* 1·414) are more meaningful here because the recording level of records is given in terms of *peak* velocity.

Peak recording levels up to at least 30cm/sec. can be detected, so if the pickup has an output of, say, 1·5mV per cm/S. recording level, then the peak output can be as high as 45mV (corresponding to about 32mV r.m.s. on a sinewave basis). To be on the safe side a 6dB (2 *times* voltage ratio) margin should be allowed, thereby bringing the figure up to 64mV r.m.s. equivalent.

The sample spec shows that this is well catered for since the overload threshold is as high as 100mV r.m.s. for 0·5% distortion level. Pickup preamplifier overload cannot be eliminated by turning the volume control down because the volume control appears at its output. Pickup overload, therefore, can be responsible for poor reproduction even when playing at moderate sound levels. The effect is not unlike pickup mistracking.

The overload margin commonly deteriorates with reducing frequency, which is another reason why a 6dB margin should be allowed. At frequencies above 1kHz the margin widens due to the action of the RIAA-equalisation on the preamplifier's feedback, but, then, at these higher frequencies the pickup output can be greater anyway!

Tape Recording Output

This is the signal output delivered by the tape recording outputs of the amplifier or hi-fi receiver. It is commonly independent of the volume control setting, tone controls and filters but dependent, of course, on the level of the source signal.

At the 'phono' output sockets the signal level is round 150–200mV 'average', which suits the majority of tape machines with this sort of recording input socket. The recording output to the DIN socket is less than this, for DIN-based tape machines call for a much smaller signal for full recording level. The DIN 'constant current' recording standard is nominally 1mV for each 1kΩ of load resistance presented by the tape machine — so if the recording input load of the tape machine is, say, 47kΩ, then the signal across the DIN socket will be round 47mV 'average'.

Other Parameters

These, then, are the 'standard' engineering parameters that accompany an amplifier or the amplifier section of a hi-fi receiver (formerly called tuner-amplifier). They do let you know how powerful the amplifier is, how well it has been designed, whether it is worth its money, etc. They *do not*, however, let you know exactly how the amplifier will audition to the listening room.

In addition to the basic engineering parameters other factors are here involved, including the quality of the signal sources and loudspeakers, how well the loudspeakers interface the amplifier and various other objective/subjective parameters which are only recently being brought to light. Some of these are considered in Chapter 9.

You must be able to judge the engineering parameters, nevertheless, to discover whether the amplifier is powerful enough for your requirements, whether it will satisfactorily match the signal sources, whether the tone control and filter characteristics are to your liking, etc.

It is sad that amplifiers are judged in terms of watts of power output. Signal voltage output across specified load conditions would be far more appropriate. This is because the sound pressure level (SPL) produced by a loudspeaker is generally rated in terms of signal *voltage* across it. A loudspeaker spec may include the parameter of 96dB SPL for 6V input measured under anachoic conditions (e.g., non-reverberant room) at 1m.

In a normal domestic situation you will be using two loudspeakers for stereo, the room will be reverberant, depending on its size and furnishings, and you will not be sitting at 1m from the loudspeakers. In an 'average' listening room at normal listening distance from the loudspeaker 96dB SPL will be evoked when the signal voltage to each loudspeaker is a little over 2 *times* that of the exampled loudspeaker parameter.

In other words, each loudspeaker will require an input of about 13V. If the loudspeakers each have a nominal 8Ω impedance (equivalent resistance), then to produce 13V across each one, each channel of the amplifier will have to be rated round 21W.

This applies to a typical lounge size of some 70^3m. An SPL of 96dB is remarkably 'loud', and under most normal conditions you would generally be peaking below this. You will have to make sure your amplifier choice provides this sort of output at low distortion.

CHAPTER THREE

TUNER SPECIFICATIONS

RADIO TUNERS ARE COMPOSED of two main sections: that which processes the 'radio' (or r.f.) signal and that which handles the subsequently extracted audio (or d.f.) signal. Although the first section can have some (subtle) influence on the audio signal passed to the partnering hi-fi amplifier, the second section is generally more exacting in this respect.

R.F. Section

The first section mainly determines how well (or otherwise) the tuner will behave under different reception conditions: for example, how well it will receive weak signals for a stated background noise ('hiss') and hence signal-to-noise (S/N) ratio. It also determines how well it will accommodate weak signals from the aerial in competition with a multiplicity of strong ones simultaneously present; also how well it will separate a weak signal from a stronger one on a near frequency (selectivity). Most things like this are handled by the 'radio' section.

A.F. Section

After the 'radio' signal has been processed, the resulting weak audio signal needs to be amplified to a level suitable for application to the radio or tuner input of the hi-fi amplifier. The stereo multiplex signal, also present with the 'mono' audio signal at the detector output of the first section, is decoded in the second section. This is handled by the *stereo decoder*, which delivers separate left and right channel audio signals.

The second section of a stereo tuner thus has an audio amplifier for each channel. Other things also happen in the second section, including the required f.m. de-emphasis and the filtering of unwanted sub-channel spurious signals which are inevitably produced during the decoding process. Distortion, stereo separation, frequency response and to some extent the noise background on a stereo programme are thus influenced by the parameters of the second section of an f.m. tuner.

Difficult Reception Areas

Provided the 'signal' conditions where a tuner is operated are not unduly 'difficult' or abnormal, then it is possible to achieve valid hi-fi reproduction from a tuner whose first section is below the ultimate requirements for very 'difficult' reception areas; that is, of course, provided the parameters of the second section satisfy the hi-fi requirements.

However, such a tuner may not give very good results from 'weak' aerial signals, particularly in areas where the tuner is expected to work in competition with very strong signals from nearby transmitters. Nevertheless, for good quality reception in many areas it is not always essential to spend out on a tuner whose first section is engineered to the pinnacle of the art.

If you are interested in receiving only the 'local' stations at top quality, then look for a tuner with good second section parameters. On the other hand, if you are in a 'difficult' reception area or like tuning stations outside the service area you will also need to study the first section parameters.

Sadly, parameters of tuner specifications rarely make clear whether they are related to the first or second section. What we propose to do, therefore, is to divide the list of tuner parameters into two sections, the first relating to the 'radio' performance and the second to the audio or 'sound' performance. Such a parameter list is given below.

'Radio' Parameters

Sensitivity
for 30dB S/N mono: 1·5μV
for 50dB S/N mono: 5μV
for 50dB S/N stereo: 50μV

Limiting −1dB: 2μV

3rd-order Intermodulation: 68dB

Equivalent front-end selectivity: 46dB

Repeat spot suppression ratio: 86dB

IF rejection ratio: 90dB

Image response rejection ratio: 70dB

Capture ratio: 2dB

Selectivity
±200kHz: 5dB
±400kHz: 60dB

Audio Parameters

S/N ratio 1mV
mono: 70dB
stereo: 67dB

AM rejection ratio: 60dB

Hum: 70dB

Pilot tone rejection ratio: 50dB

Distortion 100% modulation
mono: 0·2%
stereo: 0·4%

Frequency response ref. 1kHz
20Hz: −3dB
15kHz: −1dB

Stereo separation
1kHz: 35dB
10kHz: 25dB

Distortion on breakthrough signal: 20%

Birdies interference rejection ratio: 60dB

Some of these may appear to be rather technical; but don't despair! You may also see some parameters on our list which are not included in the manufacturers' specifications. We shall explain these.

Sensitivity

Tuner sensitivity is stated as the strength of the aerial signal required for a given background noise ('hiss'). With an f.m. tuner, as the aerial input signal strength is increased so the background 'hiss' falls. The intensity of the 'hiss' is compared with the intensity of a single tone audio signal corresponding to full modulation level, called 100% modulation level.

Thus, if the noise is, say, ten times less than the actual signal we could say that at that particular aerial input the S/N ratio is 10:1. It is general practice, though, to translate the ratio to a dB ratio. A signal voltage (noise or audio signal) ratio of 10:1 is 20dB (see, for example, page 38 in *Pickups and Loudspeakers*). The dB ratio is better than a direct ratio because it is logarithmic and thus correlates to the way that we hear.

Sensitivity for 30dB S/N ratio, therefore, merely gives the aerial signal required for the noise to be 30dB (31.62:1 voltage ratio) below 100% modulation. To get the noise to fall to 50dB (316.2:1 voltage ratio) below full modulation a larger aerial input is required. You will notice that the aerial input required for a 50dB S/N ratio is greater for stereo than mono. This is due to the extra bandwidth and noise produced by the decoding action.

For hi-fi listening the S/N ratio should be, at least, 60dB in both stereo and mono modes, so although the 30dB and 50dB ratios give a good impression of the 'absolute' sensitivity of a tuner, you will require to feed the tuner with a greater aerial signal to cut the noise further.

The S/N ratio, later defined, lets you know how low the noise is when the aerial input is 1mV. Incidentally, 1μV (microvolt) is one-millionth of a volt and 1mV (millivolt) one-thousandth of a volt or 1,000μV. You can get tuners with a 30dB mono S/N ratio as low as 0·8μV, but the average is round 1·5μV.

The average f.m. tuner has a 50dB S/N ratio of about 5μV mono and 50μV stereo. There is generally a 10:1 voltage difference between the two modes, meaning that for stereo you need 20dB more aerial signal than for mono to provide 50dB S/N ratio. This difference diminishes with increasing aerial signal.

Limiting −1dB

If we now exclude the noise and look only at how the audio output from the tuner rises with increasing aerial signal we can assess the limiting. When the modulated aerial signal strength arrives at a certain value the audio output from the tuner no longer rises. It remains at a steady level. The value of signal required for this is called the full limiting signal.

Since the rise towards full limiting occurs very slowly to the nth degree, the aerial input to provide an audio output which is 1dB below full limiting is often quoted. This is called −1dB limiting aerial input. The average f.m. tuner has a −1dB limiting input close to the input required for 30dB mono S/N ratio.

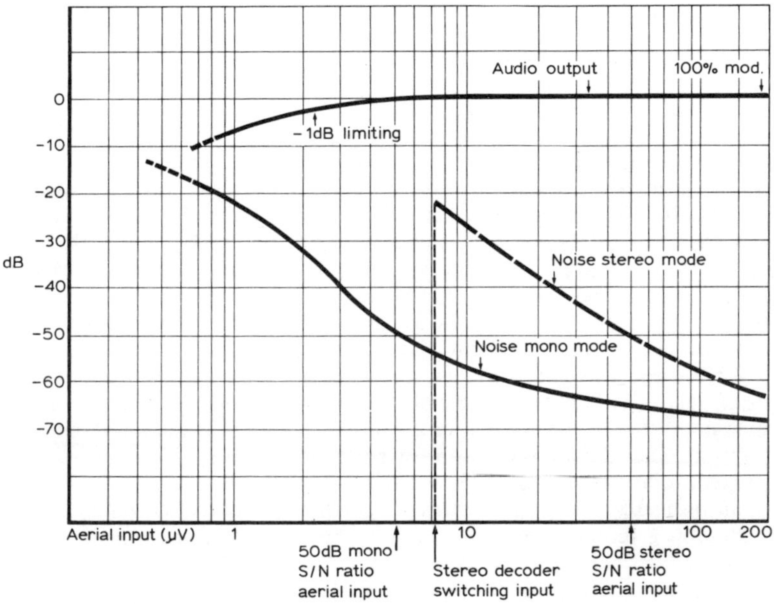

Fig. 3.1: Curves showing the parameters of limiting and mono and stereo S/N ratios.

If you are interested in receiving weak signals and in long-distance (DX) reception, where the signal is likely to be weak and fading (rising and falling in strength), then you should certainly look for a tuner whose −1dB limiting occurs at the lowest possible aerial input. If the limiting comes on at a too high aerial input, the strength of the audio signal fed to the amplifier will rise and fall in sympathy with a signal fade or the level will be different between stations of different strength.

The curves in Fig. 3.1 summarise the points so far discussed.

3rd-order IM (RFIM)

An f.m. aerial responds to all the signals present in the v.h.f. (very high-frequency) f.m. band (called Band II) and possibly other signals outside the band. It is the job of the first stages of the tuner to select the required signal. This selection happens progressively as the aerial signals pass through the front-end.

As with an amplifier, there is a limit to the strength and number of signals that can be handled by the front-end linearly. A multiplicity of relatively strong signals, depending on the large-signal handling ability of the front-end, can cause overloading. When this happens intermodulation products are generated. These are spurious signals which are not actually picked up by the aerial but which are created in the tuner circuits, and once they have been created they cannot be eliminated.

The frequencies of these signals are related to the frequencies of the signals from the aerial in a rather complex way. For example, if the tuner is receiving simultaneously signals at, say, f1 = 94MHz and f2 = 95MHz, then a spurious signal at 2f2 −f1 = 96MHz and another at 2f1 −f2 = 93MHz can result. These are called 3rd-order intermodulation (IM) products.

With more than two signals, the interactions become much more complicated. The products could, of course, interfere with the wanted signals. With the three frequencies of a station group of f2, f3 and f4, corresponding to Radios 2, 3 and 4 on, say, 89·1MHz, 91·3MHz and 93·5MHz respectively (e.g., the Wrotham transmitter), 3rd-order IM can give rise to a spurious signal at f2+f4−f3 which, if you work it out, you will find falls right on top of the Radio 3 signal at 91·3MHz. A tuner of poor 3rd-order IM performance could thus give interference on Radio 3.

When the measurement is made, the strength of two signals applied to the tuner for a 30dB interference ratio is determined and is referred to the strength of the signal required for the ordinary 30dB S/N ratio in terms of a dB ratio. Thus the greater the dB figure of the 3rd-order IM parameter, the greater the immunity the tuner has to this sort of interference.

Unfortunately, few manufacturers' specifications carry this parameter in spite of its great importance; but tuner and receiver reviews under the names of John Earl and our associated company Gordon J. King (Enterprises) Limited appraise this aspect of tuner performance by a consistently used method of measurement, allowing ready comparisons. With our measurement, an average value would be about 58dB; the ratio of a poor tuner in this respect would be round 30dB and a top-flight tuner 70dB or more.

Again, if you are interested in DX-ing and/or live in an area where the local signals are very strong, you should look for a tuner with an average, at least, 3rd-order IM dB ratio.

Equivalent Front-End Selectivity

The required signal of the multiplicity of signals delivered by the aerial is selected in the tuner front-end by variable-tuned circuits operated by the tuning control knob or, perhaps, push-buttons. Selection is never one-hundred per cent sharp. This means that when a particular station is tuned in, signals either side of it will produce a front-end response. The degree of this response at frequencies either side of the tuned frequency is determined by the overall selectivity of the front-end.

The more variable-tuned, ganged circuits there are, the better the selectivity and hence the greater the discrimination given to unwanted side-frequency signals. If the selectivity is poor, this will not necessarily mean that you will be plagued by adjacent-frequency stations breaking through to the required one.

This mode of discrimination is handled in the i.f. stages, and is appraised under 'selectivity'. What it does mean, however, is that the front-end circuits likely to produce 3rd-order IM will be receiving a higher level of signals either side of the tuned signal, thereby increasing the probability of interference due to 3rd-order IM.

Therefore, for DX-ing and for difficult reception areas it is desirable for the tuner to have a relatively high value of front-end selectivity. This is another parameter not present in the manufacturers' specifications. Our lab, however, has devised a method of determining the equivalent front-end selectivity ('figure of merit' — see *Wireless World* February 1975), and as we use this consistently in our reviews and reports you can easily compare tuners in this respect.

The more tuned circuits there are, the greater the dB number. The average value is 40dB, a relatively poor value 30dB and a top-flight value 70dB or more. It is noteworthy that the measurement also assesses the 2nd-order IM performance of a tuner.

Repeat Spot Suppression

Tuners suffer other spurious responses in addition to those already considered. One reason for this is that a superhet receiver uses a local oscillator, whose tuning is ganged to the signal variable-tuned circuits, to provide the intermediate-frequency (i.f.) signal.

Normally, the i.f. is produced by the local oscillator tracking the signal frequency as the tuning is adjusted by a frequency which is always the i.f. above the signal frequency (fs) — in a few cases the oscillator runs at the i.f. below fs. Thus, if the tuner is set to 95MHz and the i.f. is 10·7MHz (the standard for FM tuners and receivers), then the oscillator will be running at 95+10·7MHz, or 105·7MHz.

By various means, signals at frequencies other than 95MHz could 'mix' with the oscillator signal (fo) to yield the i.f. One condition is when fo has a 2nd harmonic component which, at the 95MHz tuning point, would fall at 211·4MHz. Now, if there happens to be a strong signal at 100·35MHz also arriving at the tuner, a transistor in the front-end could run into non-linearity and from this produce a 2nd harmonic at 200·7MHz.

We would then have the 2nd harmonic of fo at 211·4MHz and the 200·7MHz 2nd harmonic of the 100·35MHz signal, the difference being exactly 10·7MHz, corresponding to the i.f. The 100·35MHz signal could thus cause interference or generate a spurious response. This is called the *repeat spot* response or the half-i.f. response because the signal causing the trouble is half the i.f. (5·35MHz) away from the tuned frequency.

This is another parameter which is rarely given in the manufacturers' specifications. It is given, however, in our own reviews and test reports, measured consistently for comparison purposes. The dB ratio of this parameter should always be high, an average value being round 80dB.

A tuner whose front-end tends to overload early or whose local oscillator has a relatively high 2nd harmonic yield may return a ratio not much higher than 60dB. On the other hand, the ratio from a top-flight tuner could be greater than 100dB. The highest ratio possible, consistent with tuner price, should be selected for DX-ing and 'difficult' reception areas.

IF Rejection Ratio

This parameter indicates the susceptibility of the tuner to a signal arriving at the aerial at the i.f. It is a function of the front-end selectivity and is superseded by the more meaningful equivalent front-end selectivity measurement. Where it is found in specs, the value should be as high as possible —80dB or more.

Image Response Rejection

This is another spurious response. If we take the case of a tuner adjusted to 95MHz and the oscillator running at 105·7MHz, then the i.f. is produced also from an aerial signal of 105·7+10·7MHz, or 116·4MHz. This is the image or 'second channel' response. Again, the response is determined by the front-end selectivity — the higher the selectivity the greater the response dB ratio.

Look for a tuner with a response dB figure greater than 50dB, which is about the average for hi-fi tuners. You will also see specs where the parameter is up to 100dB, but these are applicable to the more costly models.

You will find that a tuner with a high dB figure front-end selectivity will also have a high image response dB figure. Our lab elects to measure the equivalent selectivity rather than the image response ratio.

Capture Ratio

This parameter is applicable only to the f.m. system of broadcasting. An f.m. tuner has the ability to virtually reject a signal which is a little weaker than the wanted signal, even when the two signals fall at exactly the same frequency. This is called the 'capture effect'. There is no a.m. equivalent.

The capture ratio gives an indication in dB as to how much stronger the wanted signal needs to be referred to the unwanted signal on the same frequency for a 30dB interference ratio. An average value is round 3dB. A value round 5dB would be regarded as 'poor' for a hi-fi tuner. Expensive tuners and f.m. receivers often boast a ratio as small as 1dB or less. You will see, therefore, that the smaller the dB number in this case, the better the capture ratio.

A tuner with a small dB number capture ratio is highly desirable in areas where hills, large buildings, etc. cause signal reflections. It is well known that on TV, reflected signals produce ghost images to the right of the main image. On f.m. the reflections cause harmonic distortion which increases in severity with increase in modulation; they are also responsible for impaired results on stereo.

The reflected signal (or signals) arrives a fraction of a second after the main, direct-route signal. Thus, a tuner with a good capture ratio helps to suppress the effects of the reflected signal(s).

Selectivity

The output of the front-end is at the i.f. The i.f. signal is amplified by transistors and/or integrated circuits (ICs) within a carefully controlled bandwidth. Tuned circuits and ceramic or 'surface acoustic wave' (SAW) filters at a fixed frequency provide the required selectivity (bandwidth) characteristics.

For high modulation level stereo signals the effective bandwidth needs to be round 240kHz if distortion is to be avoided. However, the side-skirts of the

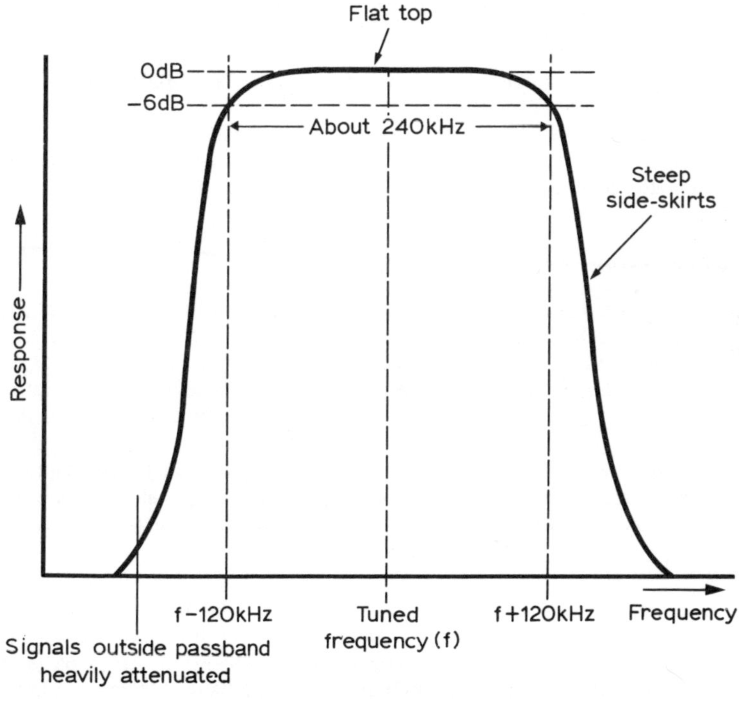

Fig. 3.2: Bandpass characteristics of i.f. channel.

bandpass shape need to be very sharp to reduce the response to side-frequency signals (see Fig. 3.2). The filters also need to be designed for small group delay for the best results.

The selectivity is evaluated at side frequencies of ±200kHz (adjacent channel) and ±400kHz (alternate channel), the results being dB numbers indicative of the selectivity ratios. The phase distortion and group delay are shown up in the audio section parameters — 100% modulation distortion, stereo separation, etc.

The average alternate channel selectivity ratio is round 50dB for a hi-fi tuner. Tuners with less exacting i.f. stages may have ratios down to about 30dB, while top-flight designs sometimes boast ratios as high as 80dB or more. The adjacent channel selectivity ratio is obviously much smaller, sometimes being less than 4dB or even a negative value (based on a 30dB interference ratio using two test signals).

A high selectivity ratio is required for DX-ing and for use in areas where relatively powerful signals are present in adjacent or alternate channels (a channel width is 200kHz) referred to the wanted signal. However, if the bandwidth is cut too much in favour of higher selectivity ratios, the sound quality can suffer.

Some tuners are equipped with switchable bandwidth, the narrow setting and correspondingly high selectivity ratio being used for seeking out weak stations close to powerful ones and the wider setting for the best quality reproduction of the 'local' service area stations.

A tuner with an alternate channel ratio round 50–60dB has a fairly reasonable all-round application.

That, then, concludes the 'radio' section parameters. You will have noticed that some of these reflect into the audio side; but shortcomings in this respect are highlighted in the audio parameters themselves, which we shall now study.

S/N Ratio 1mV

To a certain limit, the greater the aerial input applied to an f.m. tuner, the greater becomes the S/N ratio. A fairly typical signal strength is 1mV, so it is logical to measure the mono and stereo S/N ratios with an aerial input of this level. Again, the signal part of the ratio corresponds to 100% modulation level (e.g., ±75kHz deviation mono and ±67·5kHz audio deviation *plus* ±7·5kHz pilot tone and sub-carrier deviation stereo).

The stereo ratio is always less than the mono ratio because the actual audio of a stereo signal is about 1dB less than the full modulation audio of a mono signal and also because of the extra noise produced by the stereo decoding.

Noise weighting

Before the noise is put into the S/N ratio it is subjected to a 'weighting'. Noise components are wideband, and the components in certain parts of this wideband are more subjectively annoying than components in other parts. It would be illogical, therefore, to include all the wideband components in the S/N ratio.

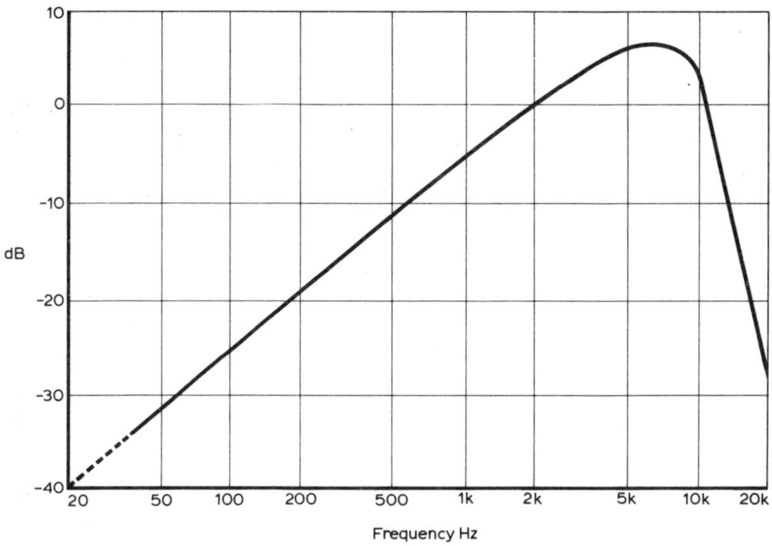

Fig. 3.3: CCIR/ARM weighting characteristic.

We all know, for example, how irritating a two-stroke motor bike engine can be. On the other hand, the noise produced by a powerful car engine is far more acceptable; yet the wideband noise readings of the two might well be identical! In general, middle- and upper-middle-to-high-frequency noise is more irritating than low-frequency noise and hum. Very high-frequency noise, of course, is not audible, anyway.

The weighting, therefore, constitutes a filter which gives more prominence to the noise components (frequencies) which annoy most and less prominence to those which annoy least.

A variety of networks have been evolved over the years for noise weighting, and that favoured by our lab is known as the CCIR characteristic. For the evaluation of broadcast and 'professional' equipment the OdB reference of the curve is set to 1kHz and a quasi-peak responding meter used to measure the noise (CCIR recommendation).

CCIR/ARM Characteristic

For the evaluation of consumer equipment, such as hi-fi, the reference is sometimes set to 2kHz and an average responding meter (ARM) used to measure the noise. This is known as the CCIR/ARM characteristic (Fig. 3.3). This configuration is recommended by the Dolby labs, who are working to establish it as a standard.

The difference between the 1kHz and 2kHz reference is essentially one of zero calibration only. With any given meter, the S/N *ratio* measured with the

2kHz reference will be approximately 6dB greater than that measured with the 1kHz reference. The type of meter also affects the noise reading. For example, an ARM reads about 11% lower on gaussian noise than a true root mean square (r.m.s.) meter, so the S/N ratio measured with the former type of meter will be about 1dB higher than the ratio measured with the latter type meter.

Using the CCIR/ARM characteristic, the average mono S/N ratio at 1mV is about 74dB and the average stereo ratio about 67dB. Tuners with mono and stereo ratios respectively as high as 83dB and 76dB have been measured. A stereo ratio much less than 62dB with this weighting would be pretty poor. In the stereo mode, the tuner's output also includes pilot tone signal and residual sub-channel signals. These must be filtered out separately before the stereo noise is read.

The curves in Fig. 3.4 show how the mono and stereo S/N ratios tend to increase slightly with increasing aerial signal above 1mV. It should also be noted that noise weighting is used to assess the S/N ratios of amplifiers (see Chapter 2).

AM Rejection Ratio

F.M. tuners and receivers should be essentially unresponsive to a.m. signals, and as impulsive interference is of an a.m. nature, f.m. equipment should be bothered less by electrical interference than a.m. equipment. This also applies

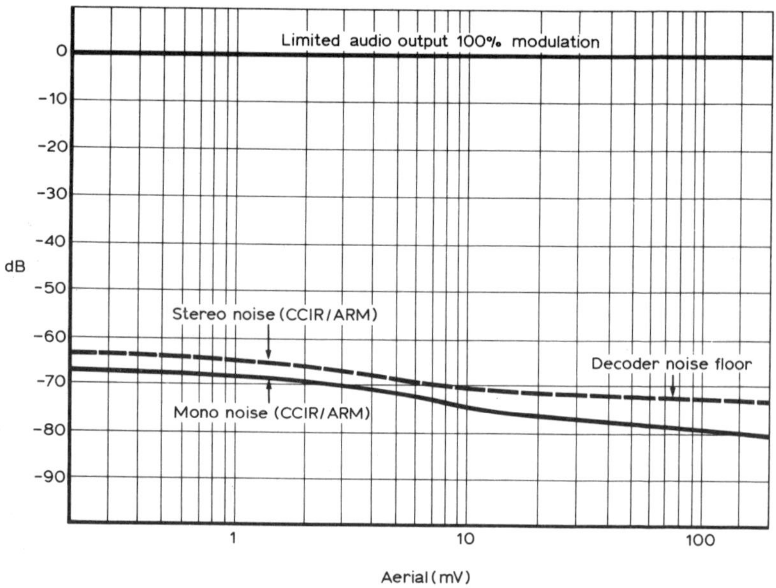

Fig. 3.4: Mono and stereo S/N ratios (CCIR/ARM weighted) at high aerials inputs, showing the 1mV S/N ratios.

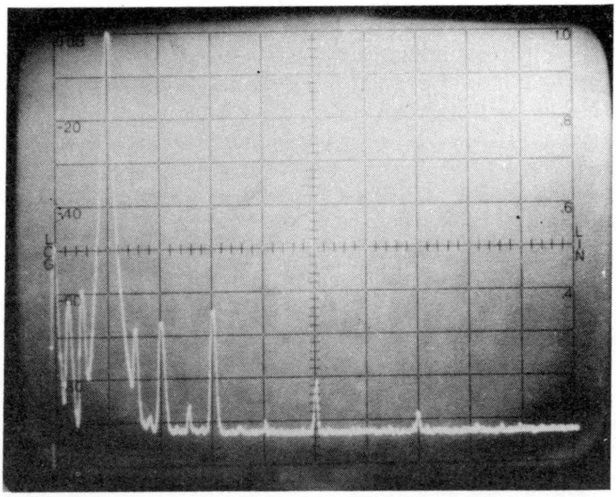

Fig. 3.5: Distortion spectrogram showing hum components at the left of the 200Hz 100% modulation signal. Scale 10dB/div. vertically and 200Hz/div. horizontally.

to 'beats' (whistles) of two signals. This, indeed, is one of the aspects of the capture effect.

The a.m. rejection ratio is usually measured with the aerial signal amplitude modulated to a depth of 30%, this then being referred to the 100% f.m. output. The measurement is made on constant tone modulation of 400Hz or 1kHz. The average tuner has a ratio round 60dB when the aerial input is 1mV. A poor value would be 40dB and an excellent value 70dB. Unfortunately, the immunity to impulsive interference may be less in practice than promised by the constant tone test.

Hum

When CCIR weighting is used to measure noise, the curve in Fig. 3.3 shows that the attenuation at the mains frequency of 50Hz is greater than 30dB. This means that the noise reading does not give a true measure of the mains frequency hum or its harmonics.

A better assessment of hum is achieved by measuring the noise with the weighting switched out. However, to measure the hum completely independently of the noise requires the use of a wave or spectrum analyser. The hum fundamental and the harmonics of the hum can then also be assessed.

The spectrogram (from a spectrum analyser) in Fig. 3.5 shows a 200Hz tone corresponding to full modulation peaking at 0dB (the top horizontal line datum) and to the left of this the hum fundamental and harmonics. In this case the harmonics and fundamental are down about −60dB. The components at the right of the 200Hz driving signal are distortion harmonics.

Sometimes a tuner tends to produce more hum when it is actually tuned to a station. This is called *modulation hum,* and the test revealed in Fig. 3.5 also assesses this because, of course, the hum is being measured while the tuner is actually responding to a modulated signal.

An average tuner produces hum fundamental plus harmonics (the vector sum) of round 70dB below full modulation when it is tuned to a station (sometimes less off-tune or in the mute mode). Tuners with a total hum of more than −80dB have been measured. It is unlikely that you would be aware of hum which is more than 65dB below full modulation under normal listening conditions.

Pilot Tone Rejection Ratio

A stereo signal carries a 19kHz pilot tone signal for the regeneration of the sub-carrier in the stereo decoder. This, together with residual sub-carrier signal, accounts for about 10% of the total ±75kHz deviation (corresponding to 100% modulation level), which is ±7·5kHz deviation. In other words, the pilot tone is transmitted at some 20dB *below* the full deviation.

Now, the left and right audio channels of an f.m. tuner are each equipped with a simple *de-emphasis* network which rolls-off the upper-frequency signals at an ultimate rate of 6dB/octave. This is required because the transmitted audio signals are subjected to *pre-emphasis* which 'boosts' the upper-frequency signals also at the ultimate rate of 6dB/octave.

The frequency at which the pre- and de-emphasis take effect depends upon a *time-constant,* which is 50μS for the British and European system and 75μS. for the American system. Correct frequency response integrity occurs only when the time constants of the pre- and de-emphasis correlate (see page 104 of *Tuners and Amplifiers*).

At 50μS, the attenuation at 19kHz relative to 1kHz is round 14 to 15dB. Thus, relative to full stereo modulation the level of the pilot tone emanating from the output of an f.m. tuner without pilot tone rejection could be round −34dB. To avoid this signal from affecting the audio reproduction, particularly when recording stereo off-air, many tuners incorporate a 19kHz filter in each channel which tends to 'notch out' the pilot tone.

The degree of rejection is best assessed with a spectrum analyser, and spectrogram (a) in Fig. 3.6 shows a 1kHz modulation signal set to 0dB datum (left) and the pilot tone (right) at about 34dB *down* from the 100% modulation signal. This tuner is thus devoid of a 19kHz filter. The spectrogram (b), scaled the same as (a), shows the pilot tone 60dB below the 100% modulation signal. This tuner, then, would be equipped with a 19kHz filter.

The other components shown on the spectrograms represent distortion. A noteworthy point is that where the pilot tone is essentially unsuppressed the distortion products to the left of it are higher than those where the pilot tone is suppressed. These, in fact, are intermodulation products due to interaction between the audio signal and the pilot tone.

Thus, for this reason alone, pilot tone rejection is desirable, though it must be engineered to avoid impairment to the upper frequency response (see later). Controlled listening tests suggest that a critical panel favours the auditioning of a tuner with good pilot tone rejection against one with no rejection.

A typical value is −60dB. Some models use nowadays a special 1C which completely cancels the pilot tone signal (e.g., the Pioneer TX9500II).

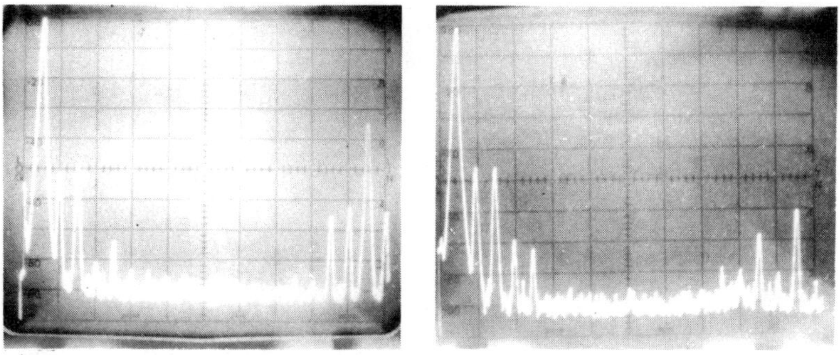

(a) (b)

Fig. 3.6: Distortion spectrograms showing amplitude of 19kHz pilot tone. (a) pilot tone at −34dB, indicating no rejection and (b) pilot tone at −60dB, indicating good rejection. Scale 10dB/div. vertically and 2kHz/div. horizontally.

Distortion

The distortion produced by a tuner depends on the level of modulation, the phase-linearity of the i.f. stages (particularly in stereo), the bandwidth of the i.f. stages (particularly at higher modulation levels), the linearity of the f.m. detector and, of course, on the design of the stereo decoder and audio stages.

The distortion is generally higher in the stereo than in the mono mode and, referred to full modulation, a good tuner should not produce much more than 0·2% total harmonic distortion mono and 0·4% stereo.

Another spectrogram is given in Fig. 3.7, with the same scaling as those in Fig. 3.6, which shows that in stereo mode the 2nd harmonic of the 1kHz 100% modulation signal is −58dB (0·126%), the 3rd harmonic −46dB (0·5%), the 4th harmonic −63dB (0·07%), the 5th harmonic at −72dB (0·025%) and the 6th harmonic at −79dB (0·011%). The vector sum of all these works out to 0·521%, which is a little higher than the average tuner. This tuner also shows the pilot tone at −40dB and quite significant IM products.

Frequency Response

Audio information carried by the f.m. system of broadcasting ranges approximately from 30Hz to 15kHz. The frequency response of a tuner, therefore, should be as 'flat' as possible over this spectrum for the best reproduction. The 19kHz notch filter (where fitted), of course, inhibits the response much above that frequency, and further upper-frequency (low-pass) filtering is used to delete sub-carrier spuriae.

A good response, even with a 10kHz filter fitted, should not be much worse than −1 to −1·5dB at 14kHz with respect to the response at 1kHz, and not much more than 1 to 2dB down at 20Hz.

The sweep oscillogram shown by the upper trace in Fig. 3.8 is scaled at 1dB per vertical division and the sweep is logarithmic from 20Hz to 43kHz. This shows the response at −0·5dB at 20Hz and −1dB at 14kHz, which is acceptable. The response then falls rapidly into the 19kHz pilot tone notch (see caption for frequency scaling).

The accuracy of response is determined by the design of the de-emphasis in conjunction with the 19kHz filtering. Some designers are able to secure better results than others.

Stereo Separation

The lower curve in Fig. 3.8 is scaled at 10dB per division vertically and shows the amount of crosstalk in the non-speaking channel (e.g., stereo separation). Here we see that the separation remains round 40dB over all the spectrum, which is excellent.

It will be understood, of course, that good separation is required for the best stereo effect. However, in addition, poor separation can lead to greater overall distortion than may be expected from the speaking channel distortion figures owing to the intrinsically high level of distortion on the breakthrough signal (see below).

'Spitting' on transients can also sometimes be heard when the high-frequency separation is poor. An average value is 35dB at 1kHz and 25 to 30dB at 10kHz.

Distortion on Breakthrough Signal

The spectrogram in Fig. 3.9 shows the distortion on the non-speaking channel of a typical f.m. tuner. The first component is the 1kHz driving signal at about −35dB, which corresponds approximately to the stereo separation. We then see the second, third, fourth, fifth and six harmonics.

The vector sum of these referred to the amplitude of the fundamental frequency breakthrough works out to something like 25%, which corresponds to the total harmonic distortion on the breakthrough signal. Hence the reason for maintaining the best possible stereo separation!

The spectrogram also shows the pilot tone at 42dB below the speaking channel 100% driving signal, which is only 8dB below the 1kHz breakthrough signal!

Birdies Suppression

When the stereo decoder is activated by the stereo signal, harmonics of the 38kHz reclaimed sub-carrier are produced. These are predominately odd-order harmonics, the 5th, for example, being at 190kHz. Now, if the tuner is tuned to a stereo signal and there happens to be another signal in the adjacent

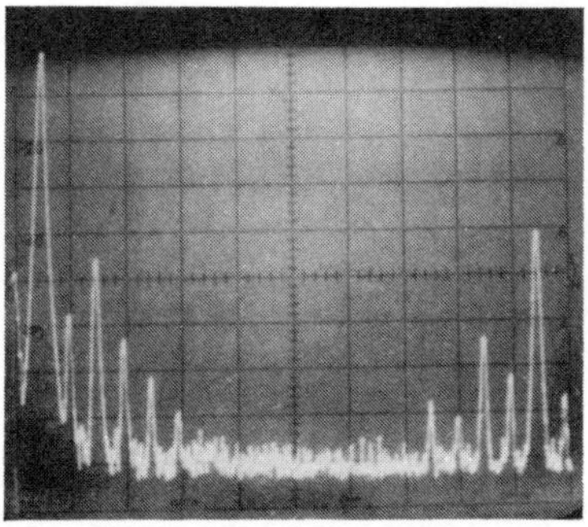

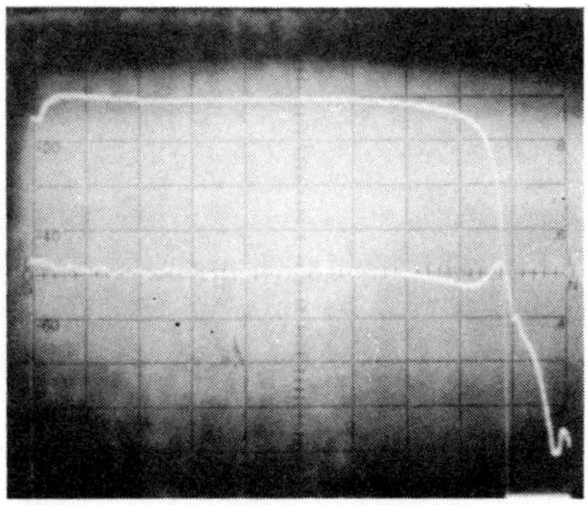

Fig. 3.7: (left) Distortion spectrogram, also showing pilot tone, where the total harmonic distortion (stereo mode) is round 0·521% (see text). Scale as Figs. (a) and (b) 3.6.

Fig. 3.8: (right) Frequency/amplitude sweeps of frequency response (upper trace and 1dB/div. vertical) and stereo separation (lower trace and 10dB/div. down from upper trace). Frequency scaling 20Hz – 43kHz logarithmic with three divisions per frequency decade.

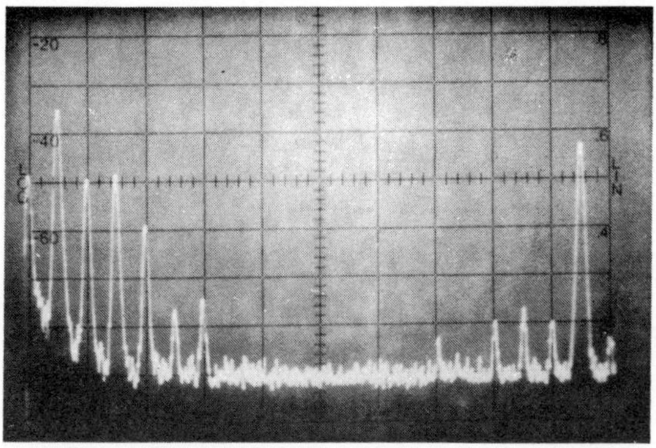

Fig. 3.9: Spectrogram showing distortion on the breakthrough signal. Relative to the driving signal in the non-speaking channel (e.g., to the fundamental of the breakthrough signal), the distortion is as high as 25%. Also showing pilot tone at –42dB relative to the speaking channel 100% 1kHz drive.

channel (200kHz away), the adjacent signal may not be heard because of the selectivity and capture effect of the tuner.

However, some of the adjacent signal might well enter the i.f. passband (see Fig. 3.2) and arrive at the f.m. detector with the tuned signal. Intrinsic non-linearity here encourages a 200kHz beat signal, which arrives at the stereo decoder. This 200kHz signal itself 'beats' with the 190kHz 5th harmonic of the sub-carrier, thereby yielding a 10kHz tone, and because this is perturbed by the modulation (frequency deviation) the 10kHz tone manifests as a 'warble', colloquially called 'birdies interference'.

How much 'birdies interference' is produced by a tuner depends on the steepness of its i.f. selectivity characteristic, on the design of the f.m. detector and stereo decoder and, of course, on the strength of the adjacent (or alternate, since there are other 38kHz harmonics present) channel signal.

Birdies rejection is never included in tuner specs, but since we rate it high in importance it is measured in our lab and is now being included in our test reports and reviews. Some tuner designers eliminate the trouble by fitting a low-pass filter (*circa* 53kHz turnover) between the f.m. detector and stereo decoder. However, unless this is done properly the high-frequency stereo separation can suffer.

Measured our way, a good value is 70dB and a poor value 50dB or less. The average value is 60dB, and with this it is unlikely whether 'birdies' would be troublesome except, perhaps, in difficult areas.

Summary

We have deliberately not ventured into a.m. radio specifications. This is a book dealing with hi-fi specifications, and the existing a.m. is far from hi-fi!

From the information presented you should easily be able to select a tuner most suitable for your particular requirements. You should be able to do this from the parameters of the spec without having to rely too much on individual 'opinion' as to the 'sounding' of a tuner!

We have found that the parameters given have a close correlation to the subjective experience of a tested, weighted and experienced listening panel working under controlled conditions.

CHAPTER FOUR

TAPE MACHINE SPECIFICATIONS

BY FAR THE MOST POPULAR tape machine at the time of writing is the cassette deck. With current design techniques allied with the new tape formulations the cassette deck has been elevated into the true hi-fi region (see *Cassette Tape Recorders* — our previous title by the publishers of this book).

The cassette deck works in partnership with a hi-fi system — providing an extra programme source while also making it possible to record from radio and disc records for future replay. The less expensive cassette recorder which contains its own amplifier and loudspeaker and the inexpensive portable machines cannot be regarded as high quality or hi-fi.

Cassette Convenience

The cassette deck is taking the place of the reel-to-reel tape deck, which was the only hi-fi tape source available a few years ago. However, there are still many enthusiasts whose interest lies in very high quality recording and replay. These buffs are more interested in the reel-to-reel deck. Certainly the quality potential of this type of machine is in advance of the cassette deck, while the wider recording tracks and the higher tape speeds greatly facilitate editing of tapes.

The convenience of the cassette system, though, is valued highly by many people, particularly where the requirement is for direct recording from one programme source — say f.m. radio or disc — for direct replay later. The compact cassette — pioneered by Philips — is far more convenient to handle and store than the disc record.

More pre-recorded cassettes are also becoming available at improved quality than hitherto with Dolby encoding for an almost 10dB noise reduction on replay when correctly Dolby decoded. To retain the advantage of the cassette while maintaining the quality potential of a wider track and faster tape speed, we have recently seen the advent of the ELCASET and similar type of cassette housings.

Parameters

The items or parameters of a tape machine are virtually the same whatever the format. What differs is the actual value of a parameter. The following example specification, therefore, could be for any tape machine — cassette, reel-to-reel or ELCASET. As we run through the parameters it will be explained how they can differ over the various formats.

Tape Deck Parameters

Input sensitivity for 0VU recording level 1kHz:
Line input: 80mV
DIN input: 0 · 8mV
Microphone input: 0 · 1mV

Input overload threshold 1kHz:
Line input: > 10V
DIN input: ≈100mV
Microphone input: 40mV

Output at 0VU 1kHz across 47kΩ: ≈400mV (output level control when fitted at maximum) from line; headphone monitoring ≈ 200mV across 8 ohms and ≈ 400mV across 200 ohms

Output overload threshold 1kHz and line: ≈ 3V

Meter reading for 200nWb/m flux: 0 to +3 VU (see text)

Harmonic (third) distortion with recommended tape:
0VU recording level: 0.3%
+6VU recording level: 2 · 5%

Intermodulation (3rd-order) distortion at 7 and 8kHz equal amplitude driving signals:
−5VU: 5%
−3VU: 8%

Head amplifier/head overload threshold: 420nWb/m flux induction for 3% 2nd harmonic distortion 1kHz

S/N ratio (CCIR/ARM — see Chapter 3) ref. 200nWb/m:
Recommended *Fe* tape: 49dB Dolby off; 59dB Dolby on
Recommended *Cr* or *FeCr* tape: 51dB Dolby off; 61dB Dolby on

Frequency response recommended tape:
20nWb/m flux: to 15kHz (−3dB)
200nWb/m flux: to 7kHz (−3dB)

Wow and Flutter (DIN peak-weighted, C90 cassette): 0 · 1%

Fast spooling time C60 cassette: 100 seconds

Stereo separation 1kHz and 0VU: 40dB

These are the parameters which our lab has determined over a number of years to be of the maximum value and which have a good degree of correlation to the listening experience. They apply essentially to a well designed cassette deck. With a good quality reel-to-reel deck some of the parameters will be better, as explained in the text which follows.

Input Sensitivity

When making a recording it is often necessary to ensure that the VU meter peaks towards 0VU on loud sounds. It is desirable for this to happen without the recording level control having to be set at maximum. The input sensitivity parameter reveals the r.m.s. signal voltage required at 1kHz to secure 0VU recording level *with the recording level control at maximum.*

The signal available for recording is generally higher in level than the input sensitivity voltage. Thus, to achieve the 0VU condition on peaks the recording level control does not need to be set at maximum.

Input Overload Threshold

However, if the source signal is very much greater than the input sensitivity voltage there is a possibility that the input preamplifiers will be overdriven on very loud sounds, causing distortion. When this is likely to happen, the recording level control will only require a small displacement from its 'zero' setting to obtain the 0VU recording level on the meters. It is highly undesirable to operate like that. The plan would then be to attenuate the input signal so that 0VU on peaks occurs when the level control is turned about half or three-quarters on.

Alternatively, one could use a different tape deck *input* whose sensitivity is closer to the source signal voltage. You will see from the example specification that the so-called 'line' input has the lowest sensitivity (requires the highest voltage for 0VU recording level). You will also see that this input is less likely to overload than the DIN or microphone input (e.g., it will accommodate a greater voltage to the overload threshold). This is the best input to use, therefore, when the source signal voltage is strong enough.

DIN Input

The DIN input is deliberately designed for relatively high sensitivity (small input voltage for 0VU). This is because it is designed to 'match' the recording output signal delivered by hi-fi receivers and amplifiers themselves equipped with a DIN socket.

The signal here depends on the input load of the tape deck presented to the output, the common DIN value being approximately 1mV for each 1kΩ of input load of the tape deck. Thus, if the load is, say, 10kΩ, the signal voltage from the DIN socket on the receiver or amplifier will be round 10mV. This will allow the recording level control to be sensibly retarded to secure 0VU recording level (the recording level control will thus be set to provide about 10 to 12dB of signal attenuation).

However, because the DIN input overloads earlier than the line input, a strong signal applied to the DIN socket is much more likely to cause distortion than when applied to the line sockets. Wherever possible, the line inputs should be used in preference to the DIN input; but where the amplifier or receiver feeding the tape deck is itself equipped with a DIN socket, then you will usually have no choice but to use the DIN socket.

Some amplifiers and receivers give a choice of line ('phono' type) sockets and a DIN socket. If the tape deck is also equipped with line outputs in addition to DIN, always use the line sockets. In this way you will be more likely to achieve the best possible recording and replay quality. The nature of the DIN socket system tends also to roll-off the higher audio frequencies earlier than the line socket ('phono') system, particularly when the screened inter-connecting leads are fairly long (see pages 154 and 159-160 of the *Cassette Tape Recorders* book).

The sensitivity of the microphone input must be very high to cater for the very small signal levels yielded by most microphones. In consequence, the overload threshold occurs earlier; but in most cases the overload margin is wide enough to avoid input overload problems, depending, of course, on the type of microphone used.

The values given for these parameters are fairly common to all deck formats.

Output Level

On replay, the tape deck provides left and right (stereo) signals of suitable level for the tape replay (or monitor) input of the partnering hi-fi amplifier or receiver. Some decks provide outputs of fixed level, ranging between 200 and 400mV. Other models include an output level control which adjusts the signal from zero up to 400mV or more (sometimes up to 1V or so).

These days an output level control is not essential because the signal level delivered by the deck is arranged to match the tape or monitor input sensitivity of the amplifier or receiver. You should thus have no problems in connecting any hi-fi deck to any hi-fi amplifier or receiver.

When the deck includes a DIN socket, the output signals (left and right channels) are approximately the same from this as from the line output sockets, which are often 'phono' type.

Output Overload Threshold

The output level previously defined is when the recording takes the meter to 0VU. Of course, when the meter goes over 0VU, the output is that much higher! Similarly, it is less at deflections below 0VU. However, it is good to know how much *output* signal can be delivered up to the point of replay output stage overload.

A typical approximate value is 3V at 1kHz. The VU meter would have to kick well against its right-hand stop before output stage distortion sets in. The head, head amplifier or tape would generally have overloaded before then! Again, these values are similar for decks of all formats.

Meter Reading for 200nWb/m Flux

Most decks are arranged so that the VU meter responds on replay as well as on record. The reading on replay is proportional to the magnetic flux recorded on the tape, and the term nanaweber per metre (nWb/m) indicates the recorded flux level (see Chapter 3 of the *Cassette Tape Recorders* book). The flux level of 200nWb/m is a convenient one because it corresponds essentially to the Dolby reference level for *cassette* machines.

With cassette machines a typical reading for 200nWb/m flux is +2 to +3VU. Few machines, however, have the reference at 0VU or a slightly different value, but not usually below −3VU. The meter reference to flux level can be important to the user, particularly when a recording is being made for the best possible dynamic range (e.g., the range between overload distortion on the one hand and noise on the other).

For example, if the 200nWb/m reference is at −3VU and the meters are peaking to 0VU, the effective flux (near but not exactly, depending on tape sensitivity) will be about 3dB above 200nWb/m, or about 280nWb/m. This means that the tape distortion is bound to be higher than when a machine is peaking to 0VU and the Dolby reference is at +3VU, for then the effective flux at 0VU will be approximately 3dB below 200nWb/m, or a mere 140nWb/m.

The converse, however, is true of the S/N ratio (see later) when this is also referred to 0VU. The ratio will be higher when the Dolby reference is at −3VU than when it is at +3VU. These days, the Dolby reference is commonly marked on the VU meters. With cassette machines remember that this corresponds to a flux close to 200nWb/m.

The Dolby reference for 6·35mm reel-to-reel tape machines is 180nWb/m; but, again, the 0VU reference on the meters may well differ from this. In fact, on domestic recorders the 0VU level is generally established with reference to a stated level of distortion (e.g., 3% third-harmonic at 333Hz).

With reel-to-reel machines the maximum record level is governed by the tape speed, being 250nWb/m at 4·75 and 9·5cm/S and 320nWb/m at 19 and 38cm/S, using 333Hz at the lower speeds and 1kHz at the higher speeds. The meter is sometimes set so that 0VU corresponds to 10dB below the maximum record level. However, when the machine is equipped with Dolby B noise reduction, the Dolby reference generally relates to 0VU. As with cassette machines, the Dolby reference is sometimes indicated on the meters. The DIN cassette reference level is 250nWb/m.

Harmonic Distortion

The amount of distortion produced by a tape machine rises in proportion with increase in recording flux. Tape distortion is predominantly third-harmonic, so it is usually this harmonic which is stated in the parameter relative either to a specific flux level or to a specified VU meter reading.

Because most users peak to 0VU on the meters, it is good to know the amount of distortion which is produced at that recording level. Accordingly, a pure (low distortion) sinewave is used to make a recording through the machine with the recording level set to 0VU. The tape is then replayed and the output sinewave is analysed (with a wave or spectrum analyser) to discover exactly how much third harmonic there is. This is then stated in the parameter either in terms of dB below the fundamental sinewave frequency or as a percentage referred to the fundamental.

A dB ratio is converted to percentage merely by finding the direct ratio from dB tables (or working it out on a pocket calculator) and dividing this into 100. For example, a dB ratio of 40dB corresponds to 100:1 direct ratio, which is thus equivalent to 1%.

Spectrogram

This test is usually made at a relatively low-frequency of 333Hz or, perhaps, 400Hz, and the spectrogram in Fig. 4.1 shows a good example. Here the horizontal divisions are 200Hz each and the vertical divisions 10dB each. In this case the driving signal (sinewave fundamental) is 400Hz and, of course, the third-harmonic 1,200Hz.

The second-harmonic is 800Hz but this is barely seen because it is below the noise floor. The spectrogram, however, does show fourth-harmonic at 1,600Hz and fifth-harmonic at 2,000Hz. The third-harmonic predominates and is at a level of approximately 30dB below the fundamental, thereby corresponding to 3%.

The amount of distortion produced is also affected by the nature of the tape (its formulation) and its biasing; also by the machine electronics and head parameters (see the *Cassette Tape Recorders* book).

Cassette machines tend to produce more distortion at a given recording flux than reel-to-reel machines using wider tape tracks and higher tape speed.

To establish how much more distortion is produced with the VU meter peaking above 0VU, a similar measurement is commonly made also at +3 or +6 VU. The example parameter values typify a good quality cassette deck operating with top-flight tape.

Intermodulation Distortion

Non-linearities at higher frequencies are best measured in terms of intermodulation distortion (IMD) by using two driving signals for recording and then analysing the adjacent sidebands (IM products).

A spectrogram showing such an analysis is given in Fig. 4.2. Here the two driving signals are at 9·5 and 10·5kHz, the first pair of adjacent sidebands thus falling at 8·5 and 11·5kHz. The two driving signals each have the same amplitude but, owing to the machine's equalisation, the sidebands differ in amplitude.

The dB ratio or percentage distortion is obtained by comparing the vector sum of the two IM sidebands with the vector sum of the two driving signals. In this case, the two driving signals show at the same amplitude at 0dB (vector sum = +3dB) on the vertical scale, while the lower sideband is −13dB and the upper sideband −15dB, corresponding respectively to 22·38% and 17·78%, the vector sum thus being 28·58%, corresponding to 20·2% distortion.

It should be noted that this is an extreme case where the recording level was deliberately set high to show the severity of this type of high-frequency distortion when over-recording. The test is sometimes made with equal amplitude driving signals of 7 and 8kHz at recording levels of −5 and −3VU, as in the example specification. It should be noted that the peak amplitude of a two-tone composite signal is 3dB above the peak amplitude of one of the driving signals. Hence the reason for making this measurement at the lower VU meter readings.

The values given in the example specification are fairly typical of a good quality cassette deck operating with a tape of the best formulation. At higher frequencies and higher recording levels the distortion can be significantly higher. Reel-to-reel machines exhibit less IMD.

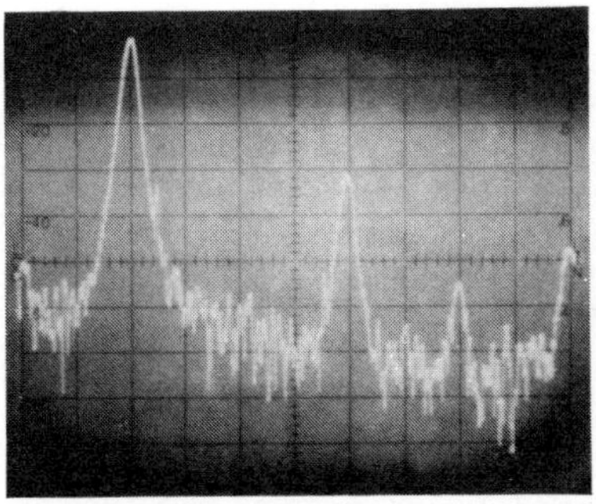

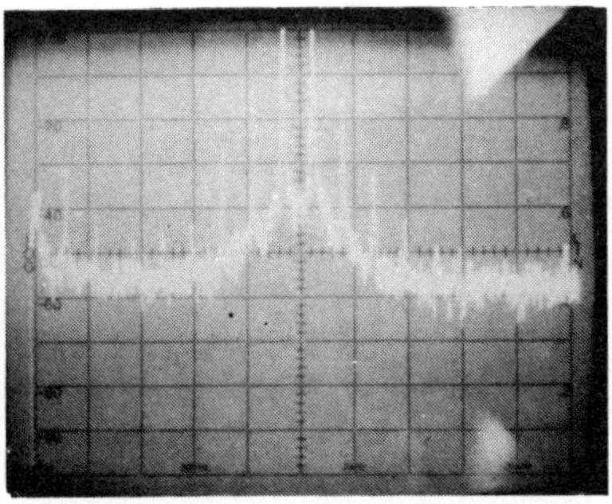

Fig. 4.1 (left): Spectrogram showing third-harmonic tape distortion. Scale 200Hz per horizontal division and 10dB per vertical division. See text for further information.

Fig. 4.2 (right): Spectrogram showing intermodulation distortion from a two-tone driving signal. Scale 2kHz per horizontal division and 10dB per vertical division. See text for further information.

Head Amplifier/Head Overload Threshold

When the signal current in the head winding, or the head signal level applied to the head input preamplifier, goes beyond a certain value, severe non-linearity sets in and distortion results. From the replay aspect, this overload threshold parameter can be important, particularly when heavily recorded tapes or musicassettes are played through the machine.

It has been suggested that the peak flux on some pre-recorded cassettes can be as high as 6dB above Dolby level, or 400nWb/m peak. Clearly, then, it is important that there is sufficient headroom to accommodate this peak flux level. If the overload threshold occurs at a lower flux value, then more distortion than necessary will occur on replay.

Similarly, when *recording* at peaks above 0VU the record head (often the same head as used for replay) can veer towards magnetic saturation and hence introduce distortion on the recorded signal. This problem is being aggravated by the new high energy, high coercivity tapes because the magnetic field required for full modulation and complete erasing is approaching the saturation limits of the heads themselves, particularly with the tapes that require the high *Cr* bias level for the best results, including the new pure iron tapes.

Some of the ferrite heads appear to be more easily saturated than those of hard permalloy. A more recent development is the *Sendust* head* which is capable of yielding high flux levels before magnetic saturation sets in. Such heads are now being found in certain JVC, Philips and other machines.

Overload Parameter

The 'overload' parameter, found in our more recent reviews and reports, is measured by applying an increasing input while analysing the output. Tape

* *This is described on page 35 of Cassette Tape Recorders*

and head distortion is essentially third-harmonic, while head amplifier distortion is generally more second-harmonic. By applying an input which raises the tape flux to 6dB above Dolby (400nWb/m) we can thus determine whether the rise in distortion is due to magnetic or amplifier overload effects.

A machine which produces no more than 3% second-harmonic at 400nWb/m would be unlikely to cause trouble under normal operating conditions. However, if the third- and other odd-order harmonics rise swiftly before this flux level, head saturation might well be troublesome.

S/N Ratio

We are now using CCIR/ARM weighting for tape machine S/N ratios referred to the 200nWb/m Dolby level for cassette machines. A new tape is recorded at the Dolby level and then, with the machine still in the record mode but with the input signal removed, the same tape is run for several minutes with no signal being recorded but, of course, with the erase system operative.

For the latter part of the process, the appropriate signal input of the machine is correctly loaded and the recording level controls set to the position corresponding to 0VU recording level when the input signal is 20dB above the input sensitivity. The measurement thus takes into account erase noise and the noise produced by the record input pre-amplifiers.

With CCIR/ARM weighting (see Chapter 3), a good *Fe* tape will give a S/N ratio of approaching 50dB with Dolby off and a *Cr* (and *FeCr* which requires 70μS equalisation) tape approaching 53dB. The ratio is improved with a noise reduction system, being about 9·5dB better with Dolby and JVC's ANRS.

Dynamic Range

Referred to the output for 3% distortion and CCIR/ARM weighting, a good machine with a good tape will have a dynamic range at 333Hz of 55dB *Fe* and 58dB *Cr* and *FeCr* without noise reduction. At high-frequency, based on IMD, ranges of 48dB and 50dB respectively can be expected. Almost an extra 10dB is achieved by noise reduction, which makes the cassette medium quite attractive for hi-fi recording and replay.

Reel-to-reel machines using wide tracks can achieve, at least, an extra 3dB dynamic range.

Frequency Response

The frequency response of a hi-fi cassette deck need be no worse than that of the f.m. system of broadcasting! A reel-to-reel machine will have an overall frequency response which is substantially better. Depending on record level, a well designed cassette deck with top-grade tape will give a response within 1dB from 30Hz to 15kHz. A reel-to-reel machine can have the same response characteristic up to 35kHz or higher.

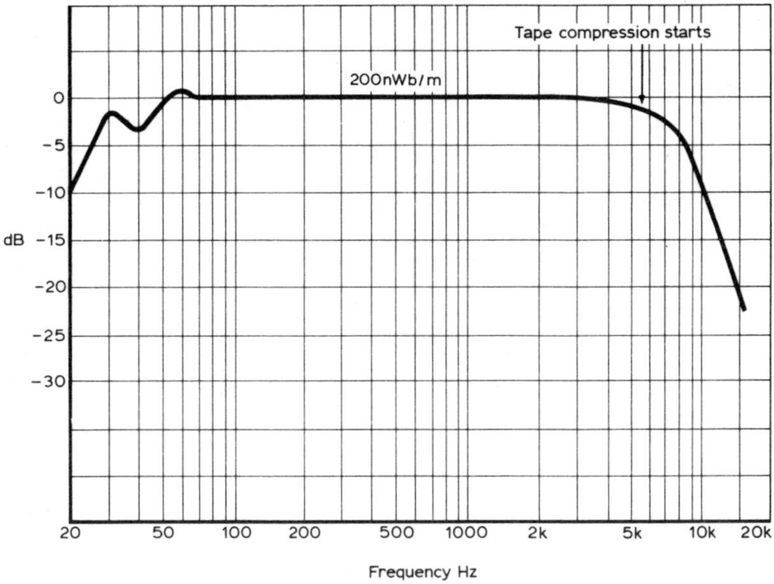

Fig. 4.3: Frequency response sweep at 200nWb/m recording level, showing tape compression.

The factor which determines the high-frequency response is essentially tape compression (assuming that the replay head is equipped with a very fine gap of not much more than a micrometer). As told in the *Cassette Tape Recorders* book, tape compression occurs progressively with rise in frequency, and it is signified by a steep rise in distortion.

The curve in Fig. 4.3 shows a frequency sweep taken at a 1kHz flux level of 200nWb/m, where it is seen that tape compression starts to show round 4 to 5kHz, the response then falling. A distortion measurement made at the compression frequency would reveal a significant rise compared with, say, the distortion at 1kHz. This is a factor of the tape itself rather than the machine.

The curve in Fig. 4.4 shows a sweep of the same machine but this time at a flux level of 20nWb/m (20dB below the first). Here the response will be seen to hold steady up to 15kHz. Just above that frequency there is a swift roll-off which, in general, is deliberately provided by a 19kHz notch filter fitted to prevent 19kHz pilot tone from poorly pilot-tone-suppressed f.m. stereo tuners from adversely affecting the operation of the Dolby noise reduction; also to prevent f.m. stereo pilot tone and tape machine bias oscillator 'beats' from causing whistles and 'burbles' on the recording.

It will be seen that this response characteristic has quite a bit in common with that of an f.m. tuner. The undulations at the bass end are caused by head characteristics, as explained in the *Cassette Tape Recorders* book.

Some people complain that such a frequency response can only be achieved at relatively low, high-frequency recording level. This is true, of course, but, in our opinion, it is not as bad as sometimes projected by equipment reviewers who seem to forget that the 'energy' of real music at high-frequency is well below the 'energy' content at the low-middle and middle frequencies, anyway!

The spectrogram in Fig. 4.5 is a slow sweep of the *1812 Overture* taken from a well recorded disc using a JVC X1 cartridge. The horizontal scale is 5kHz per division and the vertical scale 10dB per division. Notice how dramatically the effective 'energy' diminishes with increasing frequency.

Thus, when your meter is peaking towards 0VU on music signal, most of the energy of the signal contributing to this lies below the 200nWb/m tape compression onset frequency.

Wow and Flutter

Slow and fast cyclic variations of tape motion are respectively known as wow and flutter. The former can endow the reproduction with a 'drunken' effect, while the latter can impair the subtle definition and ambience of the music. Many parameters measure the two effects collectively *via* a weighting network proposed by the German DIN people (see page 116 of the *Audio Technician's Bench Manual*).

This is known as *DIN Peak-Weighted Wow and Flutter*, and is the parameter chosen for the example specification at the start of this chapter. There is reason to believe that separate readouts of wow and flutter would give more subjectively meaningful information, and we are inclined to agree with this train of thought.

The wow and flutter depends not only on the mechanical quality of the machine's tape transport, but also on the mechanics of the C-zero (the cassette itself). Some C-zeros tend to wow and flutter more than others while, on the other hand, some machines are more critical of the C-zero mechanics than others.

The parameter should assess the wow and flutter at the start, middle and end of a cassette, at the places where there can be substantial differences in torque, and the 'average' value should be given. A top-flight cassette machine will give a readout with a good cassette of less than 0·1%, while less exacting machines may go as high as 0·25% — sometimes much higher at times as the cassette torque varies and when the machine mechanics are not particularly accommodating. Much above 0·2% 'average' cannot be classified as hi-fi.

A reel-to-reel machine running at the higher speeds can yield far better wow and flutter figures than the average cassette deck. On the other hand, some of the best cassette decks return values well approaching those of reel-to-reel machines.

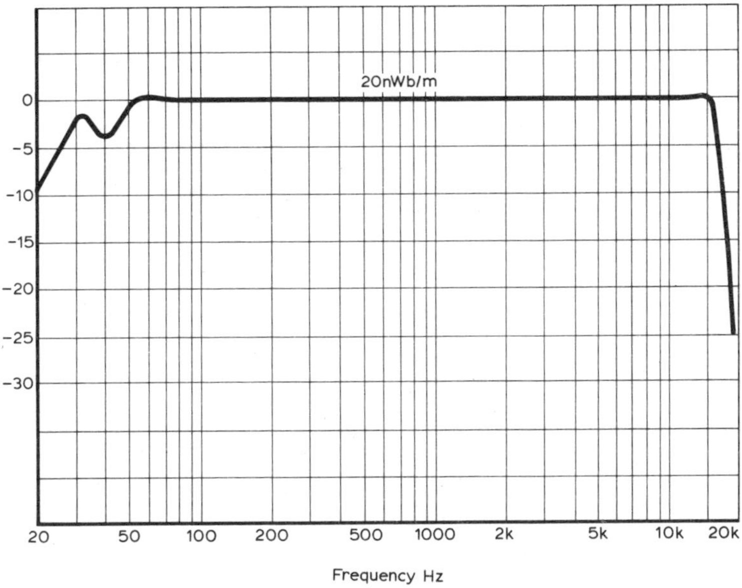

Fig. 4.4: Frequency response sweep at 20nWb/m recording level, showing 'flat' high-frequency response.

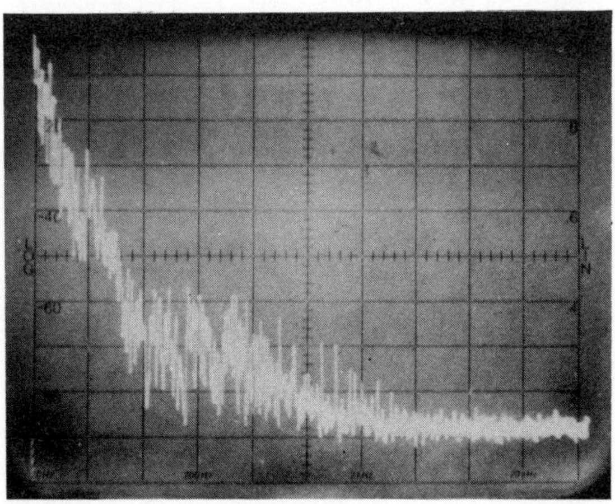

Fig. 4.5: Spectrogram of the *1812 Overture* on slow sweep, showing how the effective 'energy' of music falls off with frequency. Scale 5kHz per division horizontally and 10dB per division vertically.

Cassette	O/P 400Hz	O/P 4–5kHz	Relative Sensitiv. 1kHz	Noise	W & F	(Fig. 4.6) Frequ. response	Overall Rank
Agfa Carat**	G(–)	P(–)	P(+)	EX(+)	F(+)	(a) top G	F(+)
BASF Super LH*	EX(–)	F(–)	G(+)	G(–)	VG(–)	(a) 2nd P(–)	G(–)
Fuji FX*	G	VG(–)	G(–)	P(–)	VG(–)	(a) 3rd VG(+)	VG(–)
Grundig Ferrochrom **	F(+)	P(–)	P(–)	G	VG(+)	(a) 4th EX(+)	G(–)
Maxell UDXLI*	VG(+)	EX(–)	VG(–)	P(+)	VG(+)	(a) 5th VG(+)	EX(+)
Maxell UDXLII***	G	G(+)	EX(+)	VG(+)	VG(+)	(a) 6th G(+)	EX(+)
Philips LN*	P(–)	G(–)	P(–)	VG(+)	F(+)	(b) top VG(+)	F(–)
Philips Chrome***	F(–)	P(+)	F(–)	VG(+)	P(–)	(b) 2nd P(+)	P(–)
Pyral Maxima*	EX(+)	F(–)	F(+)	F(+)	P(–)	(b) 3rd P(–)	P(+)
Sony HF*	G(+)	VG(+)	G(+)	F(–)	EX(+)	(b) 4th F(–)	VG(–)
TDK Audua* (bias high)	P(–)	EX(–)	EX(–)	P(+)	P(–)	(b) 5th F(+)	F(–)
TDK SA***	F(–)	G(+)	VG(+)	G(+)	F(+)	(b) 6th G(+)	G(+)

Bias/equalisaton set to:
**Fe* position
***FeCr* position
****Cr* position

OdB ref. 200nWb/m recorded flux
Outputs ref. 3% distortion
Noise CCIR-weighted (machine erased tape)
Wow and flutter DIN peak weighted

P = poor; F = fair; G = good;
VG = very good; EX = excellent
Note: (+) indicates slightly better than the norm; and (–) slightly worse.
See Fig. 4.7 for marking scale.

TABLE 4.1

TAPE CASSETTE/CASSETTE MACHINE COMPATIBILITY EVALUATION
(Cassette Deck-Aiwa 1800)

Fast Spooling Time

This is merely the measure of time taken for a cassette of stated length (usually C60) to fast spool from left-to-right and right-to-left. An average time is 100 seconds for a C60 cassette.

It has been suggested that a too rapid action will tend to accelerate the oxide from the tape base when the wind suddenly halts. However, the best fast-spooling machines have automatic torque control to avoid this or other damage to the tape.

Stereo Separation

As with f.m. stereo radio, this is a measure of the signal in one stereo channel breaking into the other channel. The measurement is often made at 1kHz (sadly, with cassette decks the separation tends to diminish fairly swiftly with increasing frequency) by recording the track of one channel at 200nWb/m or 0VU, then playing the tape back and measuring the level on the signal on the non-recorded track relative to the level of signal on the recorded track.

The result is generally given as a dB ratio, and a good value is 40-50dB ref. 0VU recording level when the breakthrough signal is measured by way of a 1kHz bandpass filter to defeat noise error.

At 10kHz, the separation of a cassette machine may be as low as 20dB, depending on head and circuit design. Reel-to-reel machines can generally do better than this.

Tape Type

As will now be appreciated, the overall performance of a tape machine is strongly influenced by the type of tape used. Any specification should thus include the type of tape with which the machine's parameters were measured.

To give some idea of how the performance of a cassette machine can differ with different tapes, we show in Table 4.1 the measured parameters of the

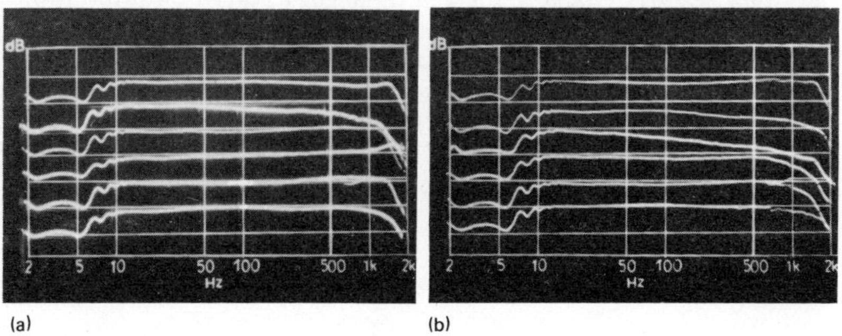

Fig. 4.6: Frequency response sweeps associated with Table 4.1, (a) first six tapes and (b) second six tapes. Scale 20Hz – 20kHz log sweep and 5dB per division vertically.

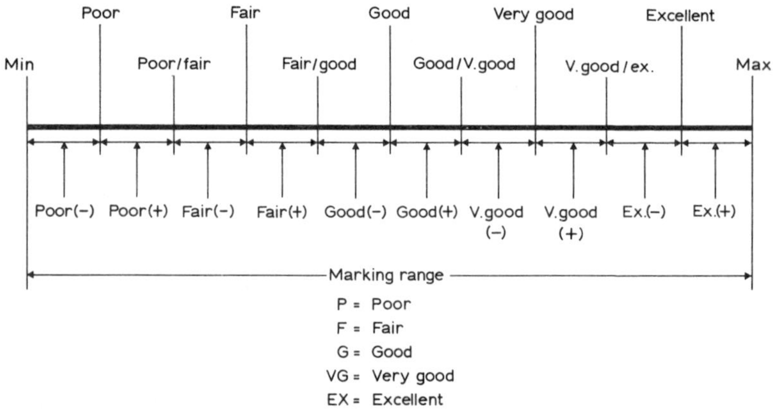

Fig. 4.7: Standard marking scale of author's lab, used for compiling the results in Table 4.1.

excellent Aiwa 1800 cassette deck with 12 different tapes*, with the frequency response sweeps at 20nWb/m given in Fig. 4.6 at (a) and (b).

The results have all been statistically analysed, but instead of presenting actual figures, we have given here the results in terms of poor (P), fair (F), good (G), very good (VG), and excellent (EX) with ± deviations according to our standard marking scale shown in Fig. 4.7.

In a practical situation, of course, one would take account of the price of the tape, for the very best results are usually obtained from the more expensive formulations. It must also be stressed that while a certain tape on one machine may give 'poor' results, on a different machine the results may be much better; which is precisely the point we are making!

Some of the tapes included have since been improved, including the Pyral formulations (with the introduction of the new *Superferrite*) and the TDK SA. Agfa have also introduced a new Fe formulation (SFD1), while BASF are currently marketing an LH1 Fe formulation and a new Cr tape. There are also a number of other excellent tapes introduced since the time of preparing this example compatability chart.

**From the work of our lab published in the Hi-Fi For Pleasure magazine, June 1977*

CHAPTER FIVE

RECORD DECK SPECIFICATIONS

A HI-FI RECORD PLAYING SYSTEM is called a *record deck.* It constitutes a programme signal source for connecting to a hi-fi amplifier and thus consists of a turntable unit mounted in a plinth — usually with a hinged transparent plastics dust cover at the top — a pickup arm and a pickup cartridge. Each item has its own specification; but there is merit in looking at the specifications of all the items together because the resulting quality of reproduction can be influenced by their compatibility — or lack of compatibility!

It is particularly important that manufacturers provide full specifications of a record deck when it does not include a cartridge. This then makes it possible for users to choose a cartridge of suitable compatibility for the arm.

Example Specs

In the example specifications which follow, the three items — turntable unit, pickup arm and cartridge — are presented separately. The subsequent text discusses, where appropriate, the aspects of compatibility and reveals the effects that the mounting plinth and dust cover can have on the reproduction. You will also find more information in this in Chapter 8.

Turntable Parameters

Features: direct drive motor, servo-control, autostop and lift and fine speed control (i.e. pitch control)
Speeds: $33\frac{1}{3}$ and 45 r.p.m.
Turntable mass: 1kg
Turntable damping: compliant mat
Time to reach correct speed: 1·5 sec. at $33\frac{1}{3}$ r.p.m.
Speed accuracy at $33\frac{1}{3}$ r.p.m.: within 0·1%
Long-term speed drift: less than 0·1%
Speed change under steady-state load: −0·2%
Wow and flutter: less than 0·1% DIN peak weighted
Rumble: −65dB DIN B weighted

Arm Parameters

Features: detachable — universal fitting — head-shell, side-thrust correction adjustment and overhang adjustment
Arm mass including headshell: 18g
Headshell mass: 7·5g
Lateral friction: 50mg

Vertical friction: 20mg
Side-thrust correction range: zero to 250mg at rim
Cartridge mass accommodation: 2 to 16g
Playing weight range: zero to 3g
Playing weight calibration error: less than 3%
Manual lifting/lowering drift/speed: none/1sec.

Cartridge Parameters

Type: moving magnet
Mass: 6g
Stylus: naked elliptical (0·0003 x 0·0007in.)
Fixing: ½-in. centres
Compliance: 15×10^{-6}cm/dyne
Effective tip mass: 0·6mg
Frequency response: 10Hz to 25kHz −2dB
Channel separation: 30dB 1kHz and 20dB 10kHz
Tracking weight range: 1·5 to 3g
Channel balance: within 2dB over response
IMD at −6dB: less than 1%
Impedance: 860 ohms/280 mH
Best tracking weight: 2·5g
Output at 5cm/sec. recording level: 5mV
Tracking threshold: pulsed high-frequency 24cm/sec. at 2g; middle and low frequencies 35cm/sec at 2g (all peak levels)

A typical specification may include other items such as dimensions, weight, clearance for dust cover, etc. On the other hand, some manufacturers' specifications may be far less detailed than our examples. Our lab measures most of the parameters exampled, for even though all of these may not directly contribute to the listening experience, they can be important in one way or another, particularly to ensure compatibility.

We shall now interpret them in turn, starting with the parameters of the turntable unit, in the order presented in the examples.

TURNTABLE PARAMETERS

Features

These merely tell you of the nature of the motor, whether direct-drive or belt-drive, whether there is servo-control for speed, pitch control (which is a fine speed adjustment) and so forth.

There are some enthusiasts who prefer belt-drive motive force in spite of the advances recently made with direct-drive motors. Chapter 8 has more information about this. Speed accuracy is often dictated by the supply mains frequency, but more expensive motors use a quartz oscillator for even greater accuracy.

Speeds

The trend is nowadays for motors to be equipped only with two speeds — 33⅓ and 45 r.p.m. If you are interested in playing early 78s, therefore, you will have to ensure that the motor provides that speed (some still do) and that you have a suitable stylus in the pickup.

Turntable Mass

This is important when the drive is *via* a belt from a relatively small motor. Some servo-controlled direct-drive (and belt drive) units use relatively low-mass turntables to secure swift reaction to load change. However, it is now being regarded as important to use relatively massive turntables even for direct-drive systems in order to iron-out minute, transient-type speed fluctuations, which are claimed in some cases to affect the reproduction, impairing the ambience and music definition.

Turntable Damping

A turntable in its 'raw' state has the unfortunate characteristic of being high-frequency resonant. That is, when tapped it will emit a bell-like ring. If a pickup stylus is placed on a stationary, undamped turntable and the output of the pickup is connected to an amplifier feeding an oscilloscope, when the turntable is tapped a decreasing amplitude sinewave (oscillation) will appear on the screen. If a loudspeaker is connected to the amplifier, the 'ring' will be heard quite dramatically through this.

When playing a record, therefore, minor turntable resonances might well be excited by the sound field from the loudspeakers impinging upon the turntable, particularly at high levels of reproduction. Clearly, this is bound to have a degrading effect on the reproducton.

A degree of damping is provided by the turntable mat. Some designs include further damping below the turntable, in the casting shell. Such resonances of the turntable and of other items of the record deck are responsible for the 'different soundings' of different record decks, including the nature and design of the plinth, the dust cover, pickup arm and ancilliary devices, such as the side-thrust correction arrangement and the lifting/lowering device.

Acoustical Feedback

Although not a parameter, this condition can be encouraged by resonances and inadequate attention to their damping. When a pickup is resting on a record and the deck is connected normally to the amplifier, a somewhat inefficient microphone situation develops. When the deck is tapped this will be heard through the loudspeakers.

Thus, when the record is playing, the sound picked up by this 'microphone' will be conveyed back to the loudspeakers through the amplifier. Owing to the resonances and the distortion on this feedback signal there can be a marked impairment in the reproduction, particularly at high sound intensities.

This effect is well known and has been with us since the advent of the record deck; but it seems to have been forgotten and is recently being rediscovered. When the sound intensity is sufficiently strong and the feedback loop gain exceeds unity, the resonances can produce a positive feedback condition. When this happens a dramatic 'howl' builds up (called *howl round*) at a frequency dictated by the various system and room resonances (including those of the pickup system and loudspeaker).

To eliminate the howl it is necessary to reduce the volume setting, switch in a low filter or retard the amplifier's bass control. A limit is thus set to the bass output and sound intensity that can be raised in a given room under these conditions from a gramophone record. If you operate at a volume setting close to the howl round threshold the reproduction is bound to be noticeably coloured.

This condition and the resulting sound coloration have been considered in our earlier hi-fi books (for example, page 186 of *Pickups and Loudspeakers,* page 153 of *ABC of Hi-Fi,* page 131 of *Improving Your Hi-Fi* and page 155 of *Audio Technician's Bench Manual,* which also proposes a method of measurement).

As mentioned in some of these references, a quick test is to place the pickup on a stationary record with the amplifier system operative, then to turn up the amplifier's volume control to find the howl round threshold setting. Turn back just below the threshold immediately, and then tap the deck plinth to discover its degree of microphony.

In general, a deck which allows a high volume control setting and one which is only mildly microphonic will audition better than one which goes into howl round at a relatively low volume control setting and which is exceptionally microphonic.

Electrical coupling from the power amplifier section back to the low-level pickup preamplifier via leads or a common impedance can also impair the listening experience. Amplifiers using separate preamp and power sections are less prone to these problems, which is one reason why they sometimes audition more favourably than integrated amplifiers.

Time to Reach Correct Speed

Unless you are a disc jockey, this is not a particularly important parameter. It is a measure of the time it takes the turntable to arrive at its stabilised speed after switch on. High torque motors can take the turntable to an accurate $33\frac{1}{3}$ r.p.m. in less than 1 second. On the other hand, you don't want to wait too long for the turntable to arrive at full speed! Direct-drive motors generally arrive at the correct speed earlier than belt-drive counterparts.

Speed Accuracy

If you are blessed with perfect pitch it is essential for the turntable to spin precisely at the recorded speed. Measurement is commonly made (in our lab, anyway) with a very accurately recorded frequency tone disc, the frequency then being read during replay on a digital frequency counter. An error of $0 \cdot 1\%$ or less is acceptable.

Many turntables are equipped with a strobe indication of speed which use the mains frequency as a reference. Others use the greater accuracy of a quartz crystal oscillator as the reference. Not all records are recorded at absolutely the correct speed and for people with perfect pitch a fine speed control (pitch control), having a range of round ±5%, would be desirable.

Long-Term Speed Drift

As you don't want to have the bother of correcting the speed at frequent intervals, once correctly set the speed should hold steady for protracted periods. Very slight speed drift is not easily detected on the strobe, but it can be a source of annoyance for people with perfect pitch.

Electronic-controlled motors appear to be more susceptible to the aberration than the basic belt-drive versions, particularly when they are equipped with a fine control for each speed. Slight drift with temperature change of components in the control section has been known to precipitate a change of up to 2% during a record playing session.

Speed Change under Load

When a normal load is placed on the rotating record there should be no significant change in speed. Some units with relatively low torque motors unfortunately exhibit a speed change when a 'dust-bug' or similar brush-type dust remover is used. This is particularly noticeable when the turntable mass is relatively low.

When electronic speed control is employed, however, the extra load results in the turn-on of more urge and, conversely, when the load is removed the urge is reduced. Sadly, this is not the whole story because when a record is

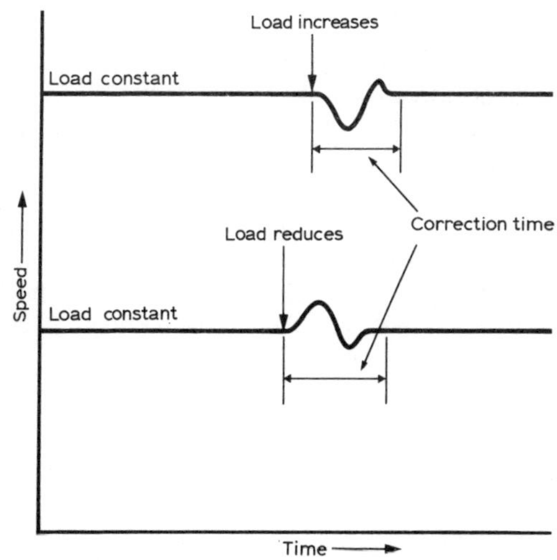

Fig. 5.1: Showing the effects of under- and over-shoot of a servo-controlled turntable unit with load changes (see text for more details).

being played the load is continuously varying owing to the changing drag resulting from the changing levels of groove modulation.

Electronic control works on the servo-controlled closed-loop principle which, by its very nature, can tend to exhibit mild under- and over-shoot effects depending on the design and the loop damping. For example, when the load suddenly increases the system might slow momentarily before the extra urge becomes effective; then when the extra power is fed to the motor it may be a little more than required, thereby giving a small overshoot before the system stabilises.

Conversely, when the load is suddenly reduced there might be a speed increase overshoot before total stabilisation. These effects are shown in elementary form in Fig. 5.1.

Thus, with continuously changing load, such as presented by a wide dynamic range record, there might be a whole complex series of under- and over-shoots which might introduce a form of very low-level, though undesirable 'modulation' on the replay signal. This could be in the form of a 'flutter' effect following steep transient recording-level changes.

It is considered by some people that this is a significant reason for one turntable unit 'sounding' different from another. It is a perfect valid observation, of course, but one which is extremely difficult to prove objectively (by measurements). It is possible to analyse flutter sidebands using a spectrum analyser with a very narrow swept filter — *circa* 1Hz, and this method, in conjunction with recorded transients, is currently being used by our lab to seek a more objectively valid way of expression.

So far, it appears that the mass of the turntable has a significant influence on the effect, the larger masses virtually ironing-out completely the tendency of such 'flutter' when the recording level changes occur quickly. Nevertheless, there is still much more we have to learn to secure a meaningful objective/subjective correlation. Right now, however, it seems that the greater the mass of the turntable, the better!

Wow and Flutter

Since this is measured on steady-state signal, such as a constant 3kHz recorded sinewave, it fails to expose the effects previously referred to. The vast majority of turntable units have highly acceptable wow and flutter figures of 0·1% or less (DIN peak weighted), and wow and flutter on steady-state signal is not subjectively detectable when below about 0·2%.

Rumble

Rumble consists of very low-frequency components of the turntable unit's bearings coupled back to the amplifier system in terms of electrical signal by the pickup. As with S/N ratios, a weighting filter is often used to emphasise the most annoying components while attenuating those which are regarded as less subjectively troublesome. A common weighting is DIN B, and provided the ratio referred to a given level of record modulation (often 10cm/S) is round the −60 to −65dB mark rumble should not cause you distress.

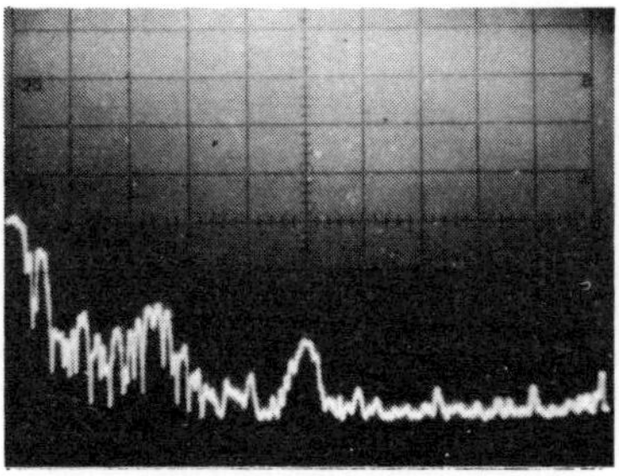

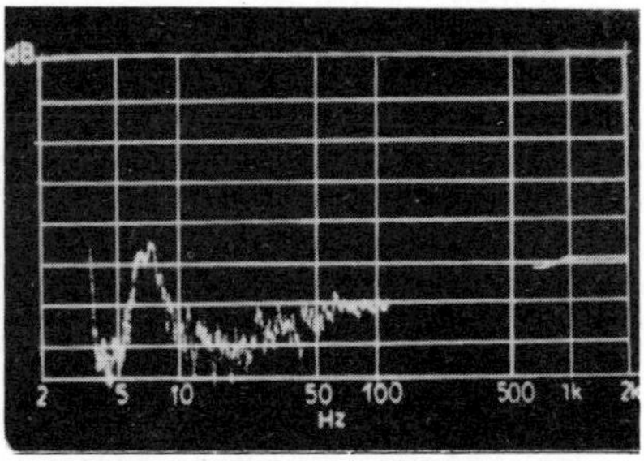

Fig. 5.2 (*left*): Rumble spectrogram of the Garrard direct-drive DD75 record deck.

Fig. 5.3 (*right*): A logarithmic sweep over 4 to 100Hz, showing the arm/compliance resonance round 7Hz.

A simple rumble parameter, however, fails to tell the whole story. More detailed reviews carry a rumble spectrogram which clearly shows the level and frequency of the rumble components below the reference signal level. Such a spectrogram is given in Fig. 5.2. The vertical scale is 10dB per division (the top 0dB horizontal line corresponding to 10cm/S recording level) and the horizontal scale 10Hz per division.

Discounting the 'starting signal', this shows a fairly heavy component round 5Hz at −55dB and a group of components at about −68dB round 25Hz. The signal at the middle of the sweep, at 50Hz, shows the amount of mains hum picked up by the deck which, at −74dB, is acceptable. This was taken from the excellent Garrard DD75 record deck, which uses a direct-drive motor.

Some direct-drive motors, particularly the earlier versions, were prone to harmonic production of the drive frequency, and in some cases components at about 100, 150, 200Hz etc. were detected at different levels. These were caused by magnetostriction effects of the rotor producing vertical vibrations of the turntable shaft.

In some cases they tended to excite turntable resonances, and for this reason some of the early direct-drive units were subjected to adverse criticism. These problems have now been largely resolved, and it needs some imagination to say conclusively that a belt-drive unit auditions better than a direct-drive unit, or vice versa!

ARM PARAMETERS

Features

Again, these tell you of the nature of the arm, the type of headshell used and whether it is detachable, and about the various adjustments. These may interest you more if you are purchasing an arm independently of the turntable unit and, perhaps, cartridge.

Arm Mass

This is an important parameter because it has a bearing of compatibility with regard to the pickup cartridge compliance. We interpret compliance later, but at this stage if you can imagine a weight (mass) supported by a spring (compliance) it is fairly easy to visualise that if the mass is moved or vibrated in some way it will wobble on the compliance. This periodic motion constitutes a resonance mode, the rate of wobble being the *resonance frequency*.

Although the pickup arm is counterbalanced and then deliberately unbalanced to provide the tracking force for the cartridge, the value of mass remains. The stylus assembly of the cartridge is of a 'springy' nature, of course, which is where the compliance comes in. Thus a resonance frequency occurs as the result of this compliance and the arm mass.

How long the wobble (mainly up and down in this case) goes on for depends on the mechanical damping, in rather the same way as the 'soft' pedal of a piano damps the vibrations (shortens the vibration time) of the strings. With a pickup system the damping may be wholly built into the cartridge or into both the cartridge and the arm. It is undesirable for the vibrations to be protracted; but if there is too much damping in the cartridge certain aspects of the tracking can suffer. Compliance *versus* damping is one of the many compromises of pickup system design.

Resonance Frequency

Vibrations, of course, are imparted to the stylus by the groove modulation, and when these vibrations correspond to the resonance frequency the electrical output at that frequency increases. At frequencies *below* resonance the output falls fairly swiftly. This means that if the resonance frequency occurs at, say, 40Hz, the low bass output of the pickup is inhibited.

There are extremely few — if any — pickup systems which exhibit a resonance due to the factors discussed at such a relatively high frequency. In general, the resonance range is between about 4 and 15Hz, only a few low-mass arm systems reaching the upper frequency, which is the most desirable end of the range.

With a cartridge of given compliance, the lower the effective mass of the arm (including headshell and cartridge), the higher the resonance frequency. We shall see later that the resonance frequency falls with increase in cartridge compliance (reduced stiffness), and because top-flight 'trackability' calls for a cartridge of fairly high compliance, the best arms are those of the lowest effective mass when properly designed. It should be remembered that it is always the stylus end of the cartridge and the record groove which has to move the arm mass, so from this aspect alone, the lower the mass the better.

Damping

When the low-frequency resonance is being excited by a corresponding-frequency modulation (or, perhaps, record warps!), the cartridge end of the arm is vibrating up and down more violently than it would without the resonance. This can affect the upper-frequency tracking because on the upward travel the downward tracking force is partially cancelled. Damping reduces the amplitude of the resonance, thereby improving the tracking, while also helping to reduce cartridge coloration.

The sweep oscillogram in Fig. 5.3 shows an arm/compliance resonance round 7Hz on a logarithmic sweep starting at 4Hz and continuing to 100Hz. The vertical scale is 5dB per division, so the resonance has a peak of some 7·5dB above the output at 100Hz. Without damping the peak would be even higher, and then there would be the probability of groove jumping if warps or record modulation (which is extremely unlikely at such a low frequency) excited the resonance.

There are various methods used for damping arms. A common one utilises a 'rubberised' arm section coupling near the counterbalance weight. Sometimes the damping is applied by a viscous fluid.

The latest Shure V15/IV cartridge is equipped with a viscous-mounted, carbon-fibre damping brush, which serves as a groove cleaner and static discharger.

Headshell Mass

The total effective mass of the arm includes the mass of the headshell and cartridge, so for a low mass system both of these items should be as light as possible. The mass of a typical detachable headshell is round 7·5g, though a headshell which is fixed or integral to the arm may have a lower mass.

Lateral Friction

Because of the currently used very low tracking forces, it is essential for the friction of the arm bearings to be as small as possible. If there is too much lateral friction the tracking will suffer badly and groove jumping may be a symptom.

The measurement is often made by applying a measured force at the stylus tip and increasing this until the arm just starts to move. This is called the 'break friction threshold force'. A top-grade arm will return a force as low as an equivalent 20mg. Provided the value is not much higher than 50mg you should be able to track high compliance cartridges without trouble (always assuming the mass of the arm is not too high).

Vertical Friction

This is measured in the same way but with the arm counterbalanced and with the force being applied at the top of the cartridge or headshell. A good value is, again, an equivalent 20mg.

Side-Thrust Correction Range

Pivoted arms are styled geometrically to minimise *lateral tracking error* by a combination of head overhang and offset. The offset angle is provided either by a straight arm and suitably angled headshell or by a curved arm. Overhang is obtained by arranging for the path of the stylus to occur a little in front of the record spindle.

The requirement is to compensate for the tracing distortion which would otherwise result from the arced path of the stylus taken by a pivoted arm relative to the pure radial cut of the recording. The technical factors involved are fully discussed on pages 64 to 67 of the *Pickups and Loudspeakers* book.

A few decks are equipped with what is usually called a 'radial tracking' pickup. With these the arm is short and straight and is arranged to track a straight line across the radius of the record. Since this path follows that of the cutter that formed the groove of the record in the first place, tracking error does not occur and so the geometric correction just mentioned is not required.

However, most arms are pivoted, and because of the geometry the pickup has a natural inclination to pull inwards while playing. This is called *sidethrust*. This tends to deflect the stylus to one side of the groove more than the other side. For a given tracking performance, this imbalance calls for a slightly greater downward force. It can also affect the balance of stereo separation between the two channels.

A hi-fi arm is fitted with a device that introduces a slight outward pull to neutralise the sidethrust. This is called *sidethrust correction*. There are three main methods of achieving this — magnetic bias, spring bias and gravity bias, the latter often using a small weight dangling on a thread.

Correction Force

The amount of correction force required depends on several factors, including the co-efficient of stylus/groove interface friction, depth of recording (i.e. recording level) and the geometry of the stylus tip. The adjustment must be made under working conditions, and even at best it can only be a compromise owing to the changing level of modulation and hence the changing drag on the stylus. Test records are available with a range of modulation levels to secure the best compromise.

The arm includes some means of adjusting the bias, which should be less at the outer diameter of a record than the inner diameter, and a typical range at the outer diameter is from zero to a maximum of about 250mg. The adjustment is eased if it can be performed while a record is actually playing. Reasonably correctly adjusted, side-thrust correction can reduce the tracking force by about 20% and improve the left-on-right and right-on-left separation balance.

Cartridge Mass Accommodation

Latter-day cartridges may range in mass from about 5-10g. When a cartridge is fitted to an arm the counterbalance weight is adjusted for accurate arm balance prior to turning on the tracking force. It is thus necessary to ensure that the arm will counterbalance accurately when the cartridge of your choice is fitted. There are not many arms that would fail to counterbalance today's cartridges.

Playing Weight Range

The arm, of course, should be capable of adjusting playing weight (more correctly, playing force) over the specified range of the cartridge to be used. However, while relatively small forces can be applied by the majority of arms, not all arms are capable of providing the forces required by relatively high tracking force cartridges.

If you use several cartridges of different tracking ranges, therefore, make sure that the arm will accommodate them all.

Playing Weight Calibration Error

The tracking force is turned on by some form of calibrated adjustment. The accuracy of the calibration is important and avoids the use of an external balance for setting the force. Most arms which earn the hi-fi label exhibit quite good accuracy in this respect. Some arms, however, have no calibration, so with these it is essential to use an external playing weight balance for the adjustment.

Lifting/Lowering Device

Many of these are damped in some way so that the downward action is slow and precise. It is also essential for accurate cueing that there is no arm drift during the lowering action. This parameter defines the rate of descent and the stability of lowering.

Maximum Lateral Tracking Error

Although this parameter is not always given, you will sometimes come up against it in arm specs. It indicates the maximum amount of lateral tracking error when the arm/cartridge combination is correctly adjusted in relation to the overhang and offset geometry over the full path of the stylus. A good arm should not introduce an error much greater than 1·5 degrees.

Although the offset angle is fixed by the design of the arm, there is provision to adjust the overhang either at the arm bass or at the headshell (slots to slide the cartridge). Adjustment optimisation usually requires the use of an alignment protractor (see page 70 of *Pickups and Loudspeakers*).

CARTRIDGE PARAMETERS

Type

There are two main types of cartridge — magnetic and piezo-electric, the latter including the so-called ceramic cartridge. Piezo-electric cartridges have the dual merits of low cost and relatively high output; but most enthusiasts would agree that a hi-fi cartridge should be a magnetic species. Other principles have been used but these have not found much favour.

There are variations of the magnetic principle, which include the moving-coil where the magnet is fixed and the coils vibrate in sympathy with the stylus, the moving-magnet where a low mass cantilever carries the stylus tip one end and a very small magnet the other end, the signals extracted from coils, the induced-magnet where a strong magnetic field is places in proximity to a low mass armature coupled to the tip working in conjunction with coils, variable reluctance which is similar to the moving armature type where the vibrating armature effectively varies the magnetic field round the signal coils, and so forth.

All the magnetic versions are capable of good performance, but the *n*th degree of performance strongly relates to the selection of the most desirable compromises and to the micro-precision of design and construction. This is why you can purchase a magnetic cartridge for less than £5, while another of similar principle might cost up to £80.

Mass

This is merely the weight of the cartridge. Today's magnetic cartridges range in mass over about 5 to 10g. Some of the moving-coil species have the higher mass value, which is reasonable because the compliance of this type is generally relatively low. The high compliance, super-tracking type, on the other hand, demand a small mass to avoid the resonance frequency (with the arm mass) from falling at too low a value.

Generally, however, the trend is towards smaller and smaller mass cartridges which is the most desirable direction of design, allied with low effective mass arms. The scheme is always to keep the moment of inertia as small as possible to minimise the 'load' carried across the disc by the stylus.

Stylus

Because the wavelengths of the sound modulation imparted in the groove are extremely small at high frequencies, and become smaller at the inner diameters of the groove spiral because the groove/stylus interface speed then falls, the active tip of the stylus must be sufficiently small to trace and define the small wavelengths with the least distortion.

Until the advent of the elliptical stylus, the tip radius was round 12·5μm (0·0005in.). Some are now being made at 17·5μm (0·0007in.). This is the so-called 'spherical' stylus. Of more recent years the elliptical stylus has virtually become a 'standard' in the better class and hence more expensive cartridges.

The elliptical cross-section is styled and fitted so that the wide dimension sits across the V-shaped groove without scraping along the bottom, while the smaller dimension provides optimum definition of the very short wavelengths. Common dimensions are 17·5μm (0·0007in.) for the major axis and 7·5μm (0·0003in.) for the minor axis.

With the coming of quadraphony we are now seeing variants, such as the Shibata stylus (Japanese), the Ichikawa stylus (also Japanese), the Pramanik stylus (B & O of Denmark) and the 'triradial' stylus (Philips). These are contoured to secure the best results from 'carrier' type quadraphonic records.

The best high-frequency tracking performance demands the smallest effective mass at the stylus tip (also see under *Effective Tip Mass*), and this can be influenced by the nature of the tip fixing to the cantilever. Some diamond tips require the intervention of metal for their fixing which, of course, adds to the mass. The more expensive cartridges are now adopting the so-called 'naked diamond' which fits directly into the cantilever without any intervening metal support.

Fixing

The vast majority of cartridges are designed to partner a standard headshell which has hole or slot fixing of ½-in. between centres.

Compliance

Compliance is a measure of the freedom of movement of the stylus. The less the stiffness, the higher the compliance. The original expression (still found in specs) is 10^{-6}cm/dyne. This merely means that a compliance of 1 (1cu) is equal to a deflection of one-millionth of a centimetre when the applied force is 1 dyne.

With the more recent SI units the expression has been changed to 10^{-3}m/N, where N is the force in newtons. When the SI units are employed mass is expressed in kg instead of in g. It all works out to the same thing, which you can discover with a little arithmetic on your pocket calculator!

A low tracking force cartridge exhibits a high compliance (good freedom of movement of the stylus); but if you use a high compliance cartridge in an arm of exceptionally high mass the low-frequency resonance will fall at an undesirably low value and the high moment of inertia of the arm will result in large deflections of the stylus when negotiating record warps. It is imperative to choose a compliance to suit the arm mass.

Effective Tip Mass

When a stylus tip is vibrating under the influence of high-frequency, high level groove modulation, high accelerations and large groove-wall forces are produced. If the effective tip mass is large, therefore, the resulting forces will impair the high-frequency tracking and damage the fine modulation patterns in the groove. The effective tip mass is not only a function of the mass of the diamond itself, but also of the mass of the cantilever and the armature, moving-magnet or moving-coil.

The micro-miniaturation of the parts attached to the cantilever and the use of light metal tubes for the cantilever are now bringing the effective tip mass down to 0·42mg (e.g., the AKG P8E and P8ES cartridges). There is still a lot of design activity going on in this area.

The recent Signet range of cartridges (by Audio Technica), for example, are now available with cantilevers of titanium, beryllium and even carbon fibre — all equipped with naked diamonds. These materials cut the mass while improving the anti-resonance properties, leading to even better translation of groove 'wriggles' to high quality sound.

Frequency Response

After correct RIAA equalisation, the frequency/amplitude response of a pickup playing a swept-tone record recorded to the RIAA standard should be flat within a dB or so from 20Hz to 20kHz. Severe undulations in response generally signify unwanted resonances. We have already looked at the very low-frequency resonance resulting from the effective mass of the arm and the compliance of the cartridge.

Another resonance occurs at the high-frequency end of the spectrum as the results of the effective tip mass and the compliance of the record material and, perhaps, the compliance of the cantilever itself. To keep this resonance well above the audio passband the effective mass of the tip must be very small — not greater than 1mg, which is on the high side of the scale nowadays.

The curves in Fig. 5.4 show at A the frequency response of a well designed cartridge where the resonances are well tamed, and at B a response signifying a cartridge of relatively high effective tip mass and poorly damped resonances. There is little doubt that the cartridge responsible for A would audition much better than that responsible for B; but it is highly likely that the first cartridge would cost a number of times more than the second!

Channel Separation

If a stereo record is recorded only in one channel, the amount of signal produced in the other channel should be very much less. This is a measure of

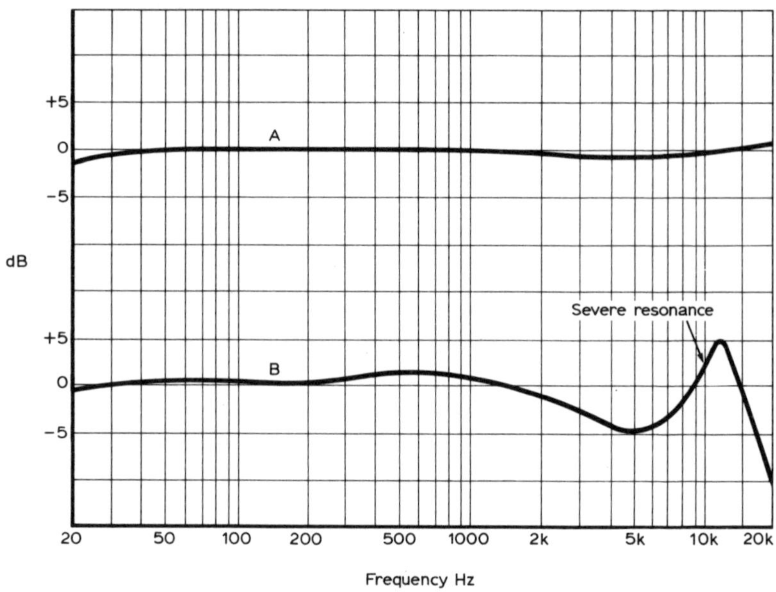

Fig. 5.4: Cartridge frequency responses. Curve A an excellent cartridge; curve B a poor cartridge with severe high-frequency resonance.

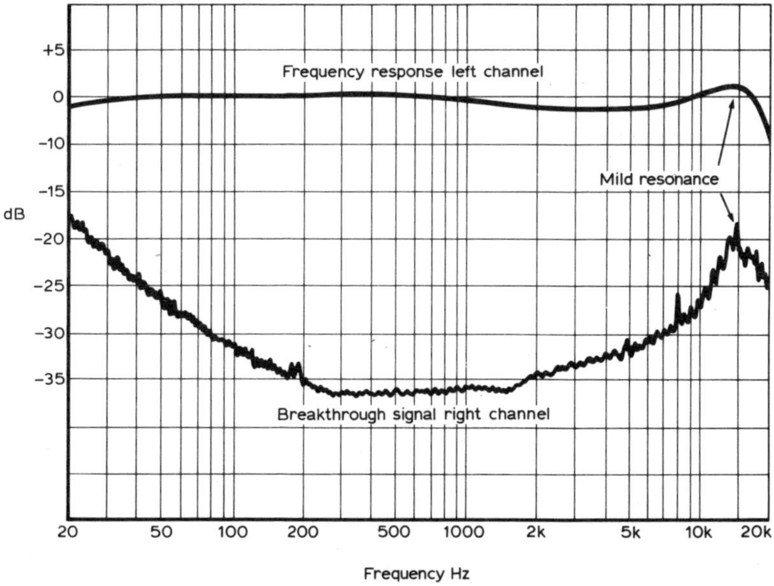

Fig. 5.5: Upper curve frequency response of left channel and lower curve the breakthrough signal in the right channel, which is an expression of stereo separation. When there is a resonance there is also usually a corresponding-frequency fall in separation.

the stereo or channel separation. A top-flight cartridge will return a separation of at least 30dB at 1kHz, with lesser separation at 100Hz and 10kHz.

A good separation is required for the best stereo effect, and stability of stereo image requires the separation to be maintained to fairly high frequencies without severe changes. Equally as important, however, is the distortion on the signal in the non-speaking channel. As with f.m. tuners, this can be astonishingly large, so if the separation is poor the reproduction will be marred by a higher than necessary overall distortion.

The curves in Fig. 5.5 show the frequency response at the top and the stereo separation below. These curves typify a good quality cartridge. Incidentally, the separation curve is also useful for showing up resonances not so easily detected on the frequency response curve, for when there is a resonance there is usually a corresponding fall in separation.

Tracking Weight Range

This is the range given by the manufacture for tracking force. In most cases the best tracking occurs at the highest force; but, unless it has been proved by

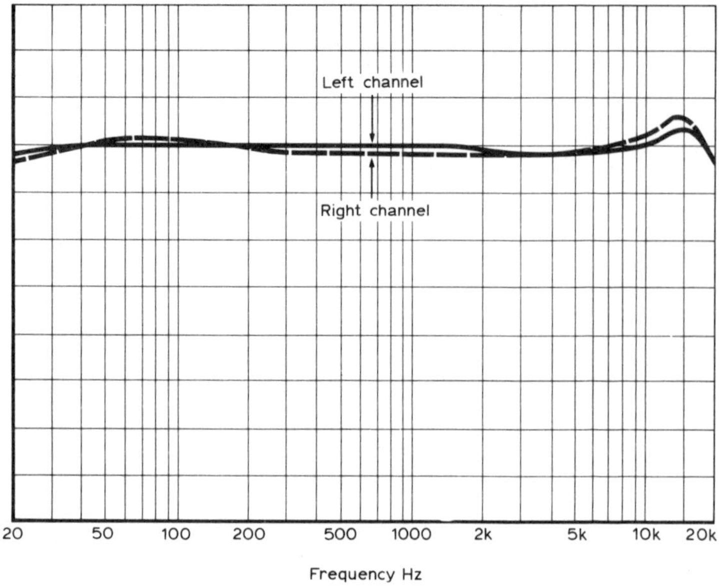

Fig. 5.6: Left and right frequency response curves, showing channel imbalance over the frequency spectrum.

independent tests otherwise, the maximum tracking force should never be exceeded. You will not improve the tracking by this means. Instead, you will increase record wear and possibly distortion.

Channel Balance

The output from a well designed and correctly adjusted pickup cartridge should not deviate very much between the left and right channels over the entire frequency range. The balance is often stated at 1kHz and a good figure is 1 to 2dB. The balance may deviate more at higher frequencies; but the smaller the change the better.

If there is a significant swing of imbalance over the spectrum the stereo image will appear to wander between the two loudspeakers. The channel balance parameter of a good cartridge over the frequency range is shown by the left and right frequency response curves in Fig. 5.6.

Distortion

The distortion produced by a cartridge or pickup system is intrinsically higher (in fact, many times higher) than that produced by a hi-fi amplifier or f.m. tuner. Most of the distortion is a function of the recording/replay geometry, and increases swiftly with recording level.

However, with due attention to the fixing of the cartridge to the arm and the subsequent fine adjustments the distortion can be kept as low as possible. Lateral tracking error, for example, can cause a steep rise in second-harmonic distortion. There is also another tracking aspect, called *vertical tracking error,* which has to do with the plane of motion of the cutter's stylus when a record is being made to that of the cartridge's stylus.

If these fail to coincide the distortion can, again, be greater than it need be. The vertical tracking angle of the cartridge, though, is fixed during design and manufacture and there is little you can do to correct this; apart from aggravating any error by incorrect fitting of the cartridge to the headshell.

Intermodulation Distortion

Sometimes the specification gives the amount of distortion to be expected at a stated recording level. This is often intermodulation distortion measured by playing a two-tone signal recording. The spectrogram in Fig. 5.7 gives you some idea of this sort of distortion from a two-tone record of 400Hz and 4kHz at 19cm/S peak composite signal recording level. The scale is 10dB per division vertically and 1kHz per division horizontally, which reveals both driving signals. The 400Hz signal is placed at the 0dB datum (top horizontal line), while the 4kHz signal is some 12dB less in amplitude.

You will see IM products either side of the 4kHz signal at about 32dB below the 0dB datum, corresponding to about 2·5%. The spectrogram also shows second- and third-harmonic distortion with respect to the 400Hz signal (at 800 and 1,200Hz) at respective levels of −34dB (2%) and −42dB (about 0·65%). Such distortions are typical of a hi-fi cartridge playing at this recording level.

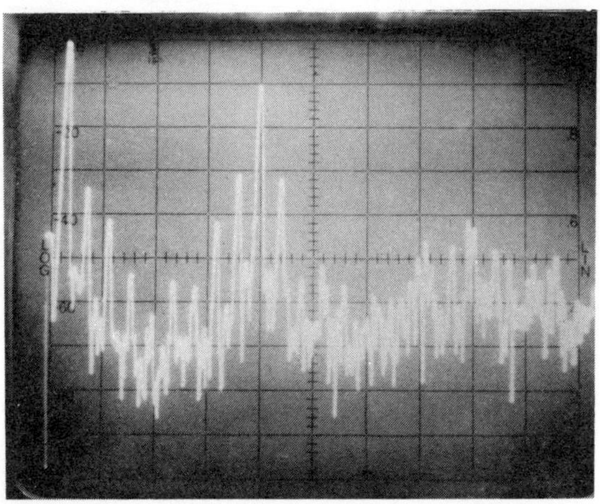

Fig. 5.7: Harmonic and intermodulation distortion of magnetic cartridge at 19cm/S peak composite signal (see text for details).

Impedance

The impedance of a cartridge is determined by the inductance of the windings and the d.c. resistance of the wire. Considering an 'average' cartridge of, say, 280mH inductance and 860Ω resistance, at 1kH the reactance of the inductance is about 1,758Ω so, with the resistance, the impedance is 1,957Ω. At 20Hz the impedance works out to about 860Ω and at 20kHz to about 35,168Ω.

A moving-coil cartridge has a much lower impedance because both the inductance and the d.c. resistance of the windings are much lower. This means that the output voltage is also much lower than that delivered by the higher impedance type of cartridge (see under *Output at 5cm/S Recording Level*).

Optimum Load

For the best overall frequency response, particularly at the treble end, a cartridge needs to be correctly loaded. This is achieved by the pickup input circuits of the amplifier. A typical load value is equivalent to 47,000Ω in parallel with about 470pF of capacitance; but, to some extent, these values are dictated by the cartridge's impedance.

Amplifier designers endeavour to arrange their pickup input circuits so that a reasonably correct load is applied to the connected cartridge, bearing in mind that some of the parallel capacitance is introduced by the capacitance of the screened leads connecting the pickup to the amplifier. Most amplifiers nowadays are arranged so that the resistance across the cartridge approximates 47,000 to 50,000Ω (that is, across each channel). This suits most cartridges; but there are problems.

For example, the capacitance of the screened connecting leads can vary between different pickup systems and record decks, while the capacitance (if any) deliberately fitted by the amplifier designer may not suit all cartridges exactly. Some amplifiers have a switch providing different values of load resistance; but only a few include a switch for adjusting the shunt capacitance.

Moreover, the pickup input of an amplifier may not be purely resistive; it may change with frequency owing to the negative feedback type of RIAA equalisation often (but not always) applied to the first, self-equalising input stages.

Feedback Conditions

Research undertaken by our lab has revealed that this changing input impedance and the interaction of certain types of RIAA feedback equalisation with the impedance of the cartridge being effectively within the feedback loop can certainly influence the upper frequency response of cartridges, some more than others.

All this is somewhat technical; but if you are interested in more detail, reference can be made to our book by Gordon J. King entitled *The Audio Handbook* (pages 51-57, published by Newnes-Butterworths) and to the work of Reg Williamson in this connection and to his article *Standard Disc Replay Amplifier* which was published in the March 1977 issue of *Hi-Fi News and Record Review*.

Certain feedback conditions occurring at low bass in some pickup preamplifiers can also impair the low bass performance of some cartridges, particularly those with a relatively high d.c. resistance. This is because with certain equalised preamplifiers the input impedance can fall dramatically at the low bass frequencies. This produces an attenuating potential-divider effect between the relatively high source resistance of the cartridge and the reduced input impedance of the amplifier. The level of the low bass signal getting into the amplifier is thus reduced.

This effect was detected and researched by Angus McKenzie while measuring numerous hi-fi receivers for his *Hi-Fi Choice — Receivers* book.

Interface Problems

The pickup/amplifier interface seems to be presenting as many problems as the amplifier/loudspeaker interface, the former which was neatly investigated by B. J. Webb in the December, 1976 issue of *Hi-Fi News and Record Review* and the latter by Gordon J. King of our lab in the same issue of the magazine.

Best Tracking Weight

This is the manufacturer's recommended tracking force with the cartridge mounted in a compatible arm and with correctly adjusted side-thrust correction. It usually falls towards the middle or upper end of the range.

It is sometimes possible to achieve improved tracking, but at the expense of slightly more record wear, by increasing the downward force a little; on no account should a force in advance of the maximum of the range be used.

Output at 5cm/S Recording Level

The electrical output of a magnetic cartridge is geared to the recording level (velocity). The greater the velocity of the stylus movement, the greater the output. The output per channel is commonly referred to 5cm/S recording level at 1kHz. On this basis the output of an 'average' magnetic cartridge is round 5mV, corresponding to 1mV per cm/S of recorded velocity. This sort of output suits the vast majority of amplifiers at the magnetic pickup input.

The output of a moving-coil cartridge is much less — 0·1mV not being untypical. This output is too small adequately to drive a contemporary hi-fi amplifier unless, perhaps, the volume control is fully advanced. Some amplifiers are now appearing with extra high sensitivity input preamplifiers to cater directly for the moving-coil cartridge, which are highly regarded and enjoying a renewed interest.

However, without this amplifier attribute, one needs to connect the cartridge to the main amplifier either through a 'booster' preamplifier (of about 27dB gain and low noise design) or a step-up transformer, which also steps up the low impedance of the cartridge to a value more suitable for the amplifier. A 50:1 voltage step-up is generally required, calling for a 50:1 turns ratio. The impedance step-up is the square of the turns ratio. Thus if the cartridge has an impedance of 2Ω, the impedance applied to the main amplifier will be 5,000Ω.

There are a few moving-coil cartridges appearing at the time of writing which have a higher output and higher impedance. These can be connected direct to the pickup input of the main amplifier without the need for amplifier 'boosting' or transformer step-up.

Tracking Threshold

A primary parameter of a cartridge is how well it will track. This is sometimes called (by Shure) 'trackability'. This can be defined as the amount of tracking force required safely to track intimately peak recorded velocities over the frequency range.

Different frequencies require different design considerations. For example, at low frequencies the amplitude of stylus vibration is greater than the velocity so a reasonably high compliance is required. At high frequencies the velocity is greater than the amplitude so a very small effective tip mass is the requirement.

At middle frequencies mechanical damping becomes more important. Mechanical damping and compliance can also interact with each other. The designer is thus faced with a multitude of compromises to create a cartridge with the best overall performance.

Tracking performance measurements are made with varying levels and frequencies of modulation and are referred to the tracking force required to handle them to the threshold of mistracking or severe rise in distortion. We have seen that the arm is a very important consideration, so the cartridge under test must be used with a high quality and compatible arm. Attention must also be given to the side-thrust correction adjustment.

Tracking Force

So far in this chapter we have been looking at tracking force in terms of downward 'weight' (hence the term 'tracking weight'). Weight, of course, produces a force as the result of gravity. It is thus not invalid to adopt the term gram (g) in this connection. For example, it is permissible to say 'tracking force 2g' or even 'tracking weight 2g'.

The early unit of force was the dyne (1,000 dyne approximately equivalent to 1g), and the unit is still used. However, with the universal acceptance of the SI (international system of) units, the newton (N) is now the unit of force (as distinct from weight which, in the SI units, is given in terms of kilogram — kg). 10,000th of a newton (e.g., 10mN) is approximately equivalent to 1g, so instead of the tracking force being stated as, say, 2g, you will find that the parameter is given in some specs as 20mN, which is more scientifically satisfying.

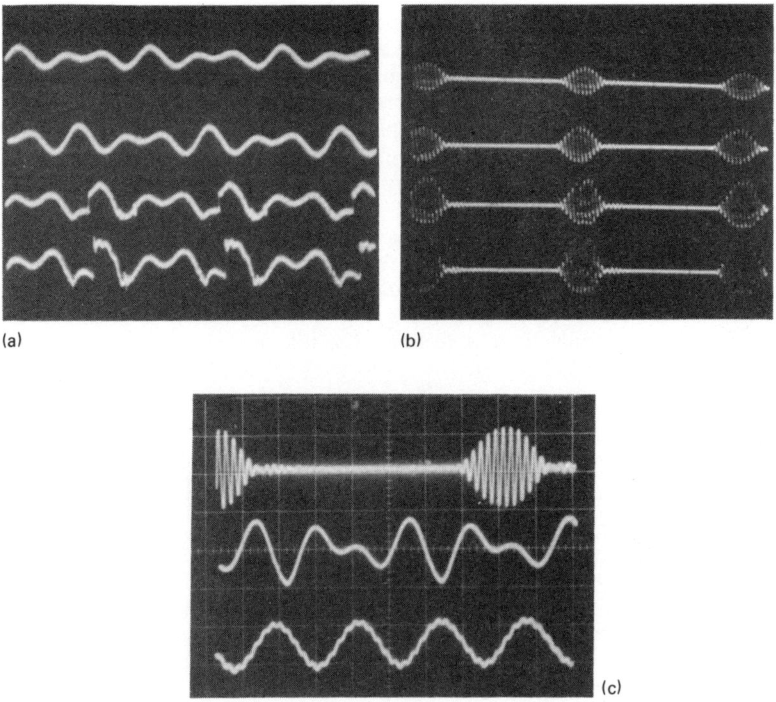

Fig. 5.8: Magnetic pickup tracking oscillograms, (a) mid-frequency, (b) high-frequency and (c) high, medium and low frequencies. See text for level details.

To correlate, of course, recording level (velocity) should really be given in metres (m) per second — m/S (the second remains an SI unit but the centimetre — cm — has been changed to m). Recording level, though, is still expressed in cm/S, which is why we have been using the term! It seems rather silly to say 24×10^{-2}m/S instead of 24cm/S.

After this discourse, you may thus find the tracking performance of a cartridge stated as 24cm/S peak at 30mN, pulsed high-frequency modulation. This merely means that a tracking force of approximately 3g is required to track pulsed-high-frequency modulation of 24cm/S peak level. This represents the threshold of tracking, where mistracking would result from an increase in recording level or a decrease in tracking force.

Tracking and Mistracking

The oscillograms in Fig. 5.8 give some illustrations of good tracking and mistracking. The two top traces at (a) show good tracking of mid-frequency modulation (1kHz and 1·5kHz composite signal) at respective levels of 20 and 25cm/S with the pickup tracking at 2g. The next two traces down show severe mistracking when the level is increased to 31·5 and 40cm/S respectively — still with the same tracking force.

Oscillogram (b) shows 10·8kHz pulsed modulation at 15, 19, 24 and 30cm/S from top to bottom, again with the cartridge tracking at 2g. Here is dramatically revealed how the mistracking increases with increase in recording level.

Oscillogram (c) shows a top-flight cartridge tracking at 1·5g 10·8kHz pulsed modulation of 30cm/S (top), 1kHz/1·5kHz modulation of 40cm/S (middle) and 400Hz/4kHz modulation of 30cm/S (bottom). All levels are peak values. Perfect tracking with no sign of breakup distortion is achieved at all frequencies, which is a reflection of a top quality, though expensive magnetic cartridge.

CHAPTER SIX

LOUDSPEAKER SPECIFICATIONS

A LOUDSPEAKER CONVERTS the electrical signal delivered to it from the amplifier to sound pressure in the listening room. Two primary parameters, therefore, are electrical input and sound pressure output.

The user also wants to know how much input is required for a given output, from which can be derived the sensitivity and efficiency, also how much input can be accommodated before the loudspeaker shows signs of mechanical and/or electrical distress or severely adds to the distortion of the output. It is also useful to know the maximum output for a given level of distortion that a loudspeaker will provide.

The input impedance, and particularly how this changes over the frequency range, can be of importance from the amplifier's point of view (loudspeaker/amplifier interface), while from the sound quality point of view we wish to know how the output changes with frequency in a controlled environment. An example loudspeaker specification is given below.

Operating principle: bass reflex
Driver units: bass dynamic 22cm; midrange dynamic 22cm; treble dynamic 5cm
Power handling: 50W amplifier
Impedance: 8 ohms nominal (see text)
Input power for 0 · 025W acoustical output: 15W
Voltage input for 96dB SPL at 1m: 10 · 25V pink noise
Frequency response: 42Hz – 16kHz ±3dB on axis
Distortion: see text

Operating Principle

This parameter usually signifies the nature of the acoustical loading used for the bass unit, and is probably related to the size of the enclosure. Most of the smaller book-shelf designs adopt the acoustical suspension type of enclosure loading — sometimes referred to as 'infinite baffle'. Larger floor-standing models may adopt bass reflex loading, where a critically-designed aperture or tunnel at the bottom exhausts the low bass sounds thereby helping with the output of the bass unit.

The nature of the loading has a bearing on efficiency. In general, bass reflex loud-speakers are more efficient than the smaller acoustical suspension designs. There are variations, including the acoustical labyrinth and tuned-pipe designs. The most efficient of all designs in the horn-loaded loudspeaker, which may have an efficiency as high as 20 or 30%, compared with the less than 1% of some of the other designs.

In general, the larger the loudspeaker the greater its efficiency and the better its low bass response, but it is noteworthy that low bass can only be reproduced adequately in a very large room. In the average listening room the long wavelengths of the low frequencies have insufficient space in which fully to develop.

Details of loudspeaker design and construction are given in our books entitled *Pickups and Loudspeakers* and *The Audio Handbook*.

Drive Units

This parameter states the number and sizes of the drive units used in the design. When there are units for bass (woofer), mid-range and treble (tweeter) the loudspeaker is sometimes called a 'three-way' design. A 'two-way' loudspeaker would thus consist of a bass unit (also handling middle frequencies) and a treble unit.

The separate units are fed with their designed-for range of frequencies from a frequency-divider network. Some designs use just one unit, called a wide-range unit; but nowadays hi-fi loudspeakers are equipped with, at least, two units.

A separate unit or driver for small frequency ranges eases the distortion problems, particularly Doppler distortion, but unless correctly designed the frequency-divider can introduce problems of its own, sometimes causing undue 'colouration' and aggravating loudspeaker/amplifier interface problems.

Power Handling

This parameter indicates the power of an amplifier that can be used with the loudspeaker when the amplifier is delivering music signal. The loudspeaker would, of course, work with lower-power amplifiers but at reduced peak sound intensity. For example, a 50W rating generally means that the loudspeaker can be safely used with a 50W per channel amplifier *provided* that the amplifier is handling music signal.

If 50W of steady-state sinewave signal were applied to the loudspeaker it might well protest, overheat and eventually fail. The average power of music signal is significantly less than the average power of steady-state signal.

However, it is less well known that an amplifier of power suitable for the loudspeaker *can* put the loudspeaker into trouble if the amplifier is driven hard so that it is clipping badly on music signal. The power in the signal then rises so that it is getting more towards the steady-state value.

This can also apply when the amplifier's rated power is *below* the rated power handling of the loudspeaker. Tweeters, too, can fail if the loudspeaker is driven fairly hard from f.m. tuner noise which occurs between stations. This is why it is a good idea to tune with the muting active. Amplifier peak clipping is another cause of unit failure.

Impedance

Since a loudspeaker system consists of resistive as well as reactive elements (i.e. the resistance of the speech coils and the inductors of the frequency-divider and the inductance and capacitance of the frequency-divider), its impedance fails to hold constant over the frequency range. The parameter thus states a nominal value of impedance which occurs round 400Hz or, perhaps, 1kHz.

Most loudspeakers exhibit a substantial increase in impedance round the 30 to 40Hz mark, corresponding to the natural resonance of the bass unit when acoustically loaded in its enclosure. At this frequency the cone vibrates vigorously (when the loudspeaker is fed from a relatively high impedance source) and the corresponding movement of the speech coil in the magnetic field generates a back e.m.f. (electromotive force) in the coil (dynamo principle), and it is this which is responsible for the impedance rise — called 'motional impedance'.

Happily, the changing impedance has less adverse effect on the reproduction than may first be thought. One reason for this is that the source impedance of a hi-fi amplifier is very low (typically between 0 · 2-0 · 1 Ω) so the loudspeaker's impedance change is 'swamped' (also see Chapter 2, particularly under *Damping Factor*), but the energy stored in the mechanical response can cause interface intermodulation distortion (IID) in the amplifier. That is, a hi-fi amplifier can be regarded as a constant voltage source, which is partly a function of the applied negative feedback.

Nevertheless, a loudspeaker is current operated, implying that the output will fall as the impedance rises and hence the current falls. At the motional impedance frequency this is not a bad thing, anyway, since at this frequency the cone is more active and tending to produce more sound output — the reduced current thus countering the effect. At higher frequencies the designer needs to keep the inductance as small as possible and retain as constant impedance as possible.

Phase Angle

The voltage across the purely resistive elements of the complex impedance is in phase with the current; but with reactance (inductance and capacitance), which comprises the other elements, this no longer holds true. The degree of phase lead or lag of the current with respect to the voltage is called the *phase angle*, which is dependent on the ratio of resistance and reactance. When the impedance is essentially reactance the phase angle will veer towards 90° and when essentially resistive the phase angle will be closer to zero degrees (in-phase condition).

Now, the amount of electrical power drawn by a load is influenced by the phase angle. When the phase angle is zero degrees, as with a pure resistance, the power is equal to the voltage across the load *times* the current flowing through it (i.e., VI); but when there is a phase angle the power drawn is equal to VI *times* the cosine of the phase angle (e.g., VIcos ø).

The term cos ø is known as the *power factor*, and when this is zero, as in the case when the phase angle is exactly 90°, the power drawn is zero watts! Clearly, then, one can never be sure just how much amplifier power a loudspeaker is consuming.

Large Phase Angle Effect

Another thing which happens when the phase angle is fairly large, and the modulus of loudspeaker impedance low, is that the output transistor protection circuits of some amplifiers are prematurely activated (for example, at a particular music frequency the current may rise to a high value when the voltage has not reached its final value).

This can cause severe distortion at an amplifier power not promised by its resistive-loaded power rating. In other words, it can severely restrict the sound pressure level (SPL), even though the amplifier may be quite powerful, relative to a given amount of distortion. More information on the phenomenon is given in Chapter 2, under *Amplifier/Loudspeaker Interface.*

We make no apologies for emphasising this aspect of amplifier/loudspeaker interface because in our judgement it is a major contribution for the different auditioning of different amplifiers when connected to different loudspeakers. Our lab has researched the problem and have discovered that certain amplifiers are more critical than others of the type of loudspeakers used with them; also, that certain loudspeakers are more difficult to partner with this type of amplifier than others (see, for example, our paper entitled Interface — Amplifier to Loudspeaker on page 87 of *Hi-Fi News and Record Review,* December 1976 issue). Also see reference to IID.

Impedance Change

The curves in Fig. 6.1 show (lower) how the impedance of a loudspeaker can change over 20Hz to 20kHz (left-hand vertical scale) and the swing of phase angle over the same frequency range (upper and right-hand vertical scale). The loudspeaker has a nominal impedance of 8Ω, shown by the broken horizontal line, and the arrow indicates a frequency where the impedance falls below the nominal at a wide phase angle.

This might incite premature operation of the protection circuits of some amplifiers. KEF is one loudspeaker firm who has been aware of this problem for many years, and who is now testing loudspeakers for phase angle as well as impedance and, of course, frequency/pressure response — issuing curves such as that in Fig. 6.1. Also see curves (a) and (b) in Fig. 6.2.

Input Power for 25mW Acoustical Output

When a loudspeaker is operating on special signal under hemispherical conditions in an anechoic chamber and the pickup microphone is placed 1m away on axis, the acoustical output of the loudspeaker is approximately 0·025W (25mW) when the sound pressure level indicated by the microphone is 96dB or 12 microbars (μb), based on the DIN technique (see *The Audio Handbook* pages 19-21).

The efficiency of the loudspeaker can thus be calculated by measuring the electrical input required for this condition. If, as in the example specification, 15W are required, then the efficiency is 0·025/50 × 100, or 0·166%, which is not untypical of a small loudspeaker of hi-fi class.

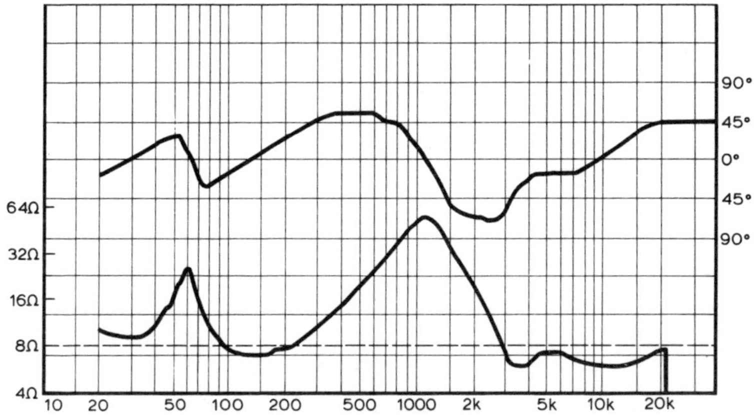

Fig. 6.1: Curves showing (top) phase angle at right-hand scale and (bottom) impedance at left-hand scale of a loudspeaker over 20Hz – 20kHz. The arrow indicates the frequency when the modulus of impedance falls below the nominal (indicated by the broken horizontal line) at a phase angle of about 45°.

Voltage input for 96dB Sound Pressure

Again, based on the DIN technique, the sensitivity of a loudspeaker corresponds to the input *voltage* of special signal required for 96dB (12μb) SPL at 1m under hemispherical anachoic conditions. This, in our judgement, is a far better way of assessing loudspeaker sensitivity, rather than in terms of efficiency based on electrical and acoustical *powers*.

We have seen that it is virtually impossible to know just how much power is being abstracted by a loudspeaker owing to its power factor. It is also much easier to tie in with amplifier output if this were given in more meaningful voltage rather than in electrical power measured across a resistive load, which is untypical of a real loudspeaker.

After all, a hi-fi amplifier is a constant voltage device and as such should be rated in output voltage obtainable across specified loads. You would then soon know whether the amplifier would (or not) drive 'difficult' loudspeakers or jib at large phase angles when accompanied by very low real parts of the impedance!

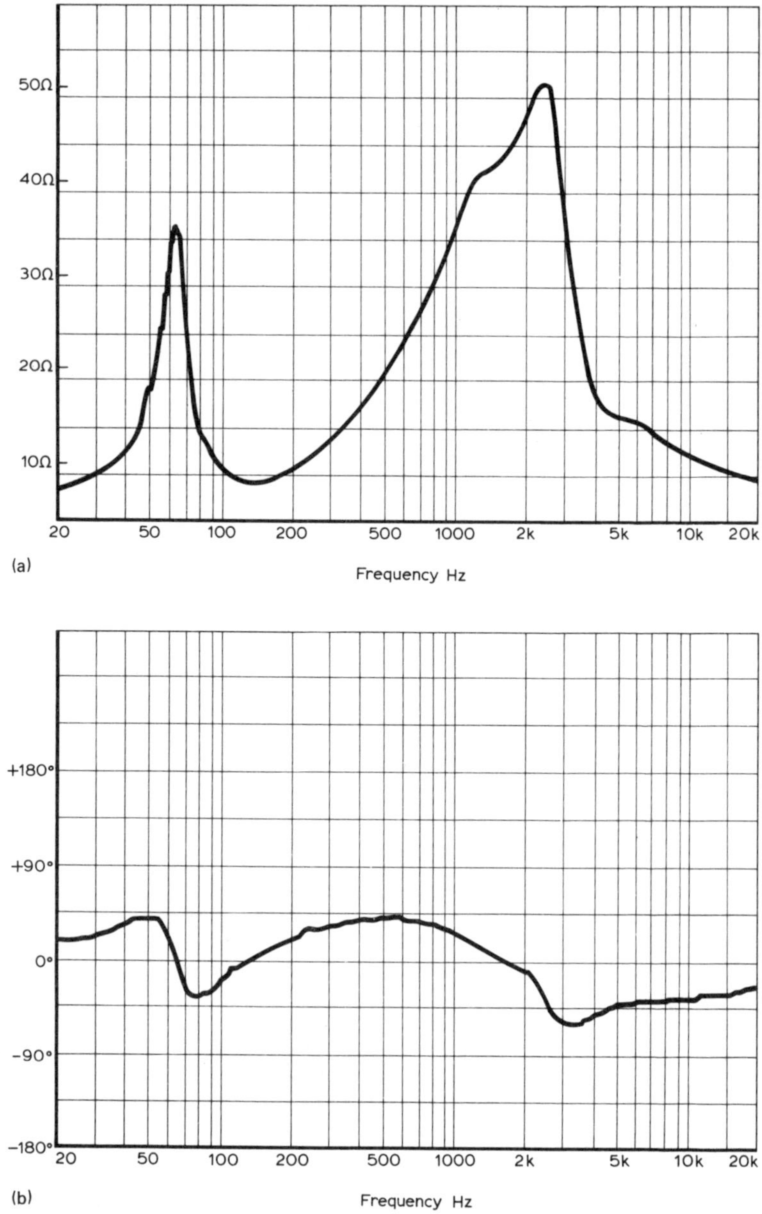

Fig. 6.2: **Loudspeaker curves, showing at (a) the modulus of impedance and at (b) the phase angle of impedance.**

Frequency Response

When terminal frequency references are given, as in the example spec, one cannot tell how the sound output pressure varies with frequency. A comprehensive spec, therefore, includes, at least, one response curve taken on axis, as the upper curve in Fig. 6.3. The lower curve shows how the response changes when the loudspeaker is turned through 30°.

Higher frequencies are propagated over smaller angles than lower ones. At very low frequencies, in fact, a loudspeaker approaches an omnidirectional characteristic, which is why the off-axis curve shows more of a difference at the high-frequency end relative to the on-axis curve. A fair amount of high-frequency distribution is not undesirable, and tweeters are designed to provide for this.

Insufficient dispersion tends to reduce the area over which the best stereo effect is obtained, while excessive dispersion, leading to the totally omnidirectional loudspeaker, has the effect of reducing the definition of the stereo image, though the overall effect is not disliked by some people.

Undulations in the response characteristic influence the auditioning of loudspeakers in different ways, depending on frequency and violence of response change. The most important part of the response is from 50Hz to 15kHz and over this range the response should be as 'smooth' as possible. However, in some designs dips are deliberately arranged to occur round the middle frequencies to counteract coloration elsewhere or to yield a specific condition. If a significant boost follows a dip some coloration is generally detected, the precise effect depending on the frequencies concerned.

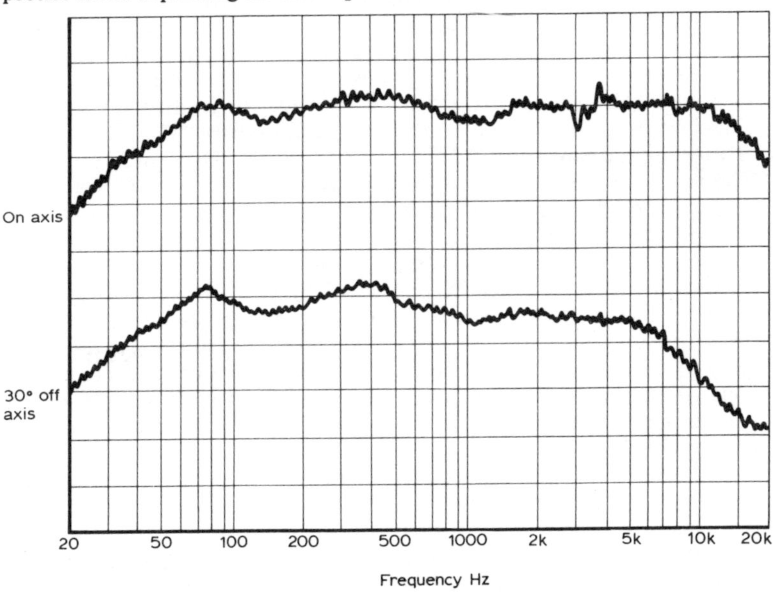

Fig. 6.3: Loudspeaker frequency response curves, upper on axis and lower 30° off-axis.

Response Characteristics

Frequency response characteristics are essential during the design stages of a loudspeaker, and they reveal how well the designer has resolved the various problems of frequency-dividing, signal distribution to the various drivers, enclosure design (at the bass end) and so forth. They cannot indicate conclusively, though, how the system will audition in the real listening room, for many of the published responses are taken under anechoic conditions (i.e., in a room without reverberation).

An entirely different response characteristic is achieved with the loudspeaker in the listening room, since then the sound reaching the ears is influenced by the reverberation times (at different frequencies) of the room, by the furnishings and the damping provided by these.

The curves in Fig. 6.4 refer to a pair of loudspeakers operating in an average listening room with both driven together, with the left only driven and with the right only driven. The input was noise signal and the response was plotted *via* a tuned third-octave filter with the sound pressure at 1kHz normalised for the different conditions. The rather violent response changes are due to resonances, a particularly bad one in the example being at 100Hz and a less violent one at 45Hz.

Distortion

Harmonic, intermodulation and Doppler distortion are often measured by the designer and, perhaps, the reviewer but are not always clearly defined in the manufacturers' specifications. The distortion rises with increase in sound output. For example, at 80Hz the 2nd and 3rd harmonic distortion of a good loudspeaker may be round 0·5% at 80dB SPL, 0·65% at 90dB and 2·5% at 100dB. At 4kHz and higher frequencies the distortion is often less, example values being 0·08% at 80dB SPL, 0·15% at 90dB and 0·2% at 100dB.

Distortion sidebands (IMD) from driving signals at 500Hz and 2kHz and a composite signal SPL of 90dB may range over 0·15% to 0·3%. Most distortion is produced at low frequencies when the bass unit is driving hard.

Doppler distortion is produced when a drive unit is called upon to handle a wide range of frequencies. For example, if the bass unit is operative at, say, 2 to 3kHz as well as at lower frequencies, the high frequencies are radiated from a cone which is also vibrating vigorously at a lower frequency.

This produces a kind of 'frequency modulation', called Doppler distortion owing to the change in high-frequency with lower frequency cone motion. Using 100Hz and 3kHz signals at a composite SPL of 90dB, Doppler distortion can range from as high as 4% down to 0·25%, depending on loudspeaker design.

It is now generally agreed, however, that hi-fi loudspeakers rarely produce sufficient Doppler distortion to be audible on music signal reproduced under normal domestic conditions.

Another distortion produced by loudspeakers and currently being unfolded is 'delayed resonance', which is shown up by impulse tests and computer analysis. The effect is that small areas of the cone tend to 'store' energy and

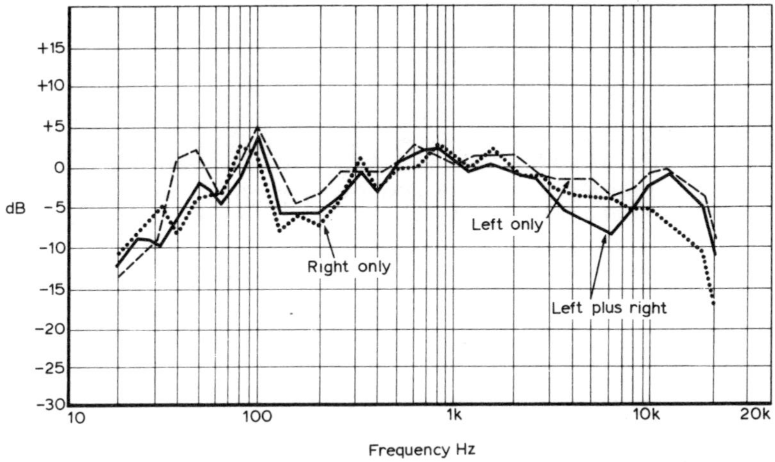

Fig. 6.4: Frequency response curves taken with a tuned third-octave filter on noise signal of a pair of stereo loudspeakers under three conditions in an average listening room (see text for details).

then release it after the signal responsible has vanished. It would appear that this is one of the most audible of loudspeaker distortions. This can cause the IID already mentioned.

Amplifier Power Requirement

Finally in this chapter, a word or two about the power requirements of the amplifier would not be amiss. A peak programme sound pressure level of 96dB in the listening room is regarded by most hi-fi enthusiasts as adequate.

To secure this sort of peak SPL from a *stereo* system operating in an average listening room you will require a power input to *each* loudspeaker which is approximately 2·5 *times* (+4dB) the power required by the loudspeaker for 25mW acoustical output. Thus with a pair of loudspeakers of efficiency equal to that of the example specification you will need a 40+40W amplifier.

Alternatively, you will need an amplifier capable of providing 4dB (1·585 times) more *voltage per channel* than the voltage input required by the loudspeaker for 96dB SPL at 1m under hemispherical anechoic conditions. Based on the loudspeaker of the example spec again, this means that you will need an amplifier capable of providing 16·25V to each loudspeaker without clipping or distorting.

CHAPTER SEVEN

SPECIFICATIONS OF ANCILLARY EQUIPMENT AND INTERFACING

MANY OF THE PARAMETERS of the specifications so far investigated are applicable also to ancillary items of hi-fi equipment. For example, the input sensitivity of a preamplifier, matrix or CD4 decoder is measured and expressed in exactly the same way at the input sensitivity of a complete amplifier though, of course, the output reference is different. The same applies to the S/N ratio.

For the correct interfacing of one item to another the output and input parameters generally need to be known. We have already seen something of this with respect to the pickup/amplifier and amplifier/loudspeaker. Interface problems can also arise with respect to the connection of a preamplifier to a power amplifier, to a decoder or graphic tone control unit between a preamplifier and a power amplifier, tuner to main amplifier or preamplifier, headphones to main amplifier, tape machine or tuner, etc. These sort of things will also be looked at in this chapter.

Hi-Fi Receiver (Tuner-Amplifier)

So far we have not referred specifically to the parameters of a hi-fi receiver or tuner-amplifier as it is sometimes called. The reason is that the parameters of the tuner section are precisely the same as those for a separate tuner and, similarly, the parameters of the amplifier section are precisely the same as those for a separate amplifer. Thus, if you are contemplating the purchase of a hi-fi receiver you can safely use as comparisons the parameters given under tuners and amplifiers.

Some hi-fi amplifiers and receivers, although being of 'integrated design' are arranged so that the preamplifier and power amplifier sections can be operated independently. Sometimes there are a couple of links connecting the two sections or a switch.

With the links removed or the switch set to 'off', the preamplifier section is disconnected from the power amplifier section. It then becomes possible to take the output of the preamplifier control section to the input of, say, a matrix decoder or graphic tone control unit (equaliser) and the output of the decoder or equaliser back to the input of the power amplifier.

In the case of the decoder, the other two outputs would be connected to a second stereo power amplifier for driving the two rear loudspeakers, as shown at (a) in Fig. 7.1. Fig. 7.1. (b) shows how a graphic equaliser would be connected.

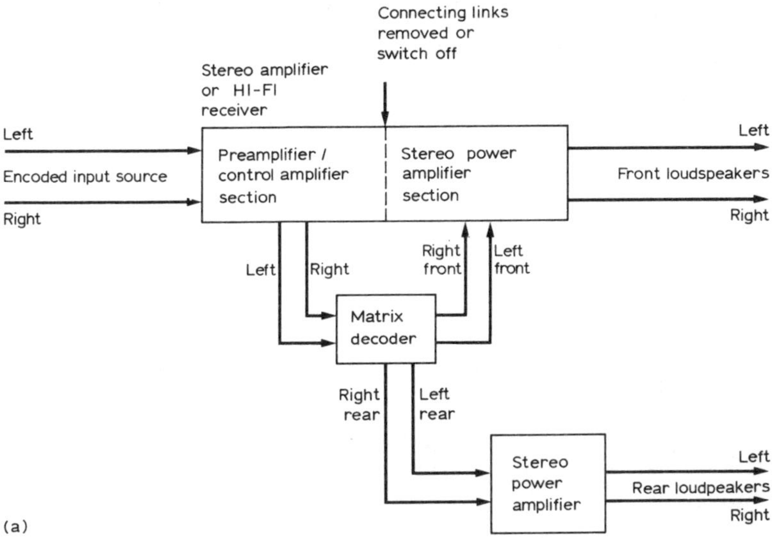

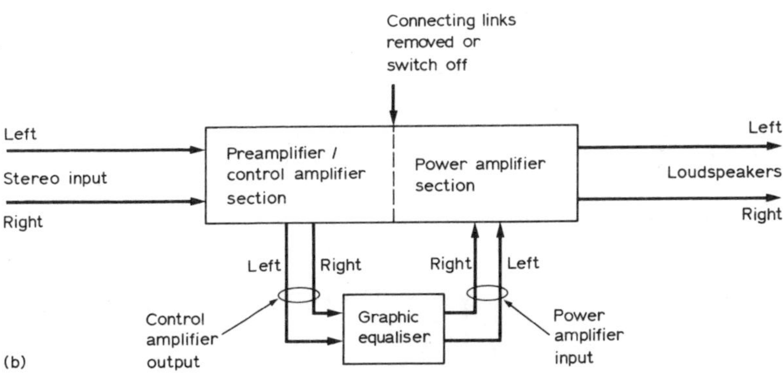

Fig. 7.1: Method of connecting matrix decoder (a) and graphic equaliser (b) to amplifier or receiver where the preamplifier/control amplifier section can be disconnected from the power amplifier section by a switch or removing a couple of links. Such an amplifier or receiver is equipped with preamplifier/control amplifier output sockets and power amplifier input sockets for the left and right channels. The diagram also reveals the connecting interfaces (see text).

Output/Input Matching

For those interconnections to work correctly, the output and input parameters must have a reasonable match. The input sensitivity of the power amplifier section is commonly around 1V r.m.s. at a fairly high impedance. Thus for full drive the decoder or equaliser must be capable of delivering, at least, this sort of output voltage.

Similarly, the input sensitivity of the decoder or equaliser must be capable of accepting the output signal level of the preamplifier/control amplifier section without running into input overload.

Input and output impedances are generally less of a problem. However, for the best results and the least likelihood of early treble roll-off due to the capacitance of the interconnecting screened leads, it is desirable for the output impedance to be lower than the input impedance.

For example, a tuner with 100Ω or so output impedance feeding into a 50kΩ amplifier input impedance is a good arrangement. To some extent the same applies to the output impedance of, say, a matrix decoder and the input impedance of a power amplifier. It is not a good thing for the input impedance to be less than the output impedance.

Matrix Decoder

It is also possible to interface a matrix decoder or graphic equaliser across the record/replay tape circuits of a hi-fi amplifier or receiver. When the tape monitor switch lies in the 'on' position a disconnection occurs between the preamplifier and power amplifier sections. Thus by using the recording outputs to feed the ancillary equipment and the output of the ancillary equipment to feed back into tape monitor similar conditions to those shown in Fig. 7.1 are obtained. The scheme is illustrated in Fig. 7.2.

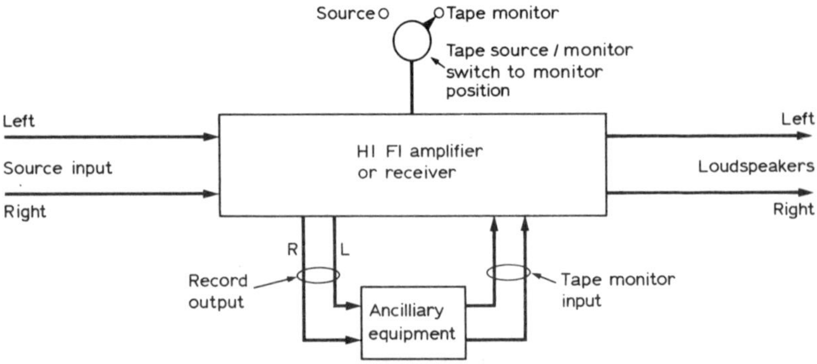

Fig. 7.2: Showing how the tape circuits can be used to interface ancilliary equipment. With this arrangement the tone controlling, filtering, etc. usually occurs after the ancilliary equipment. With the scheme shown in Fig. 7.1 it occurs before the ancilliary equipment.

This arrangement is sometimes better than the former one, but it does, of course, tie up the tape circuit. Some amplifiers and receivers, however, are now equipped with two tape circuits and switched monitoring on each, so you could use one circuit for a tape machine and the other for ancillary equipment interfacing.

It will be appreciated, of course, that when the switch is back in the source position the ancillary equipment is bypassed. When the former arrangement is used the tone controlling, filtering, etc. occurs in front of the ancillary equipment; but when the tape circuits are used it usually occurs (on replay) after the ancillary equipment.

Fig. 7.3: Showing the socketry at the rear of a Rotel receiver, Model RX-1603.

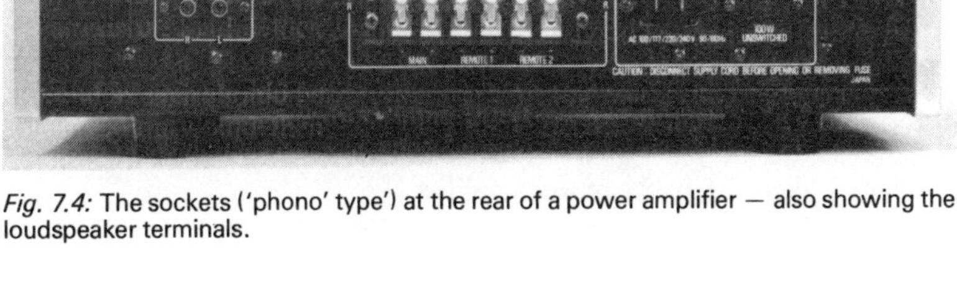

Fig. 7.4: The sockets ('phono' type') at the rear of a power amplifier — also showing the loudspeaker terminals.

Tape Circuit Interface

An advantage of using the tape circuits is that the nominal level of signal at the ancillary equipment interfaces is generally less than that when the equipment is introduced between the preamplifier/control amplifier and main power amplifier sections. For example, the tape recording signal may be round 200mV and the tape monitor input sensitivity of a similar value at the 'phono' type sockets. This lower level can be more compatible with the input and output parameters of the ancillary equipment.

You should be aware of DIN socket problems, though; the recording output from the DIN socket is commonly fed through a high resistance to provide a so-called constant current characteristic to the DIN recording input of the tape machine, the signal voltage thus being dependent on the machine's input impedance.

By using the DIN socket, therefore, you may find that there is insufficient signal fully to drive the ancillary equipment. It is unlikely that you will encounter problems at the monitor input because the DIN socket input sensitivity is the same as, or close to, that of the 'phono' type socket input sensitivity.

Separate Control and Power Amplifier Units

Although most hi-fi amplifiers combine the control and power amplifier sections in a common housing (as, of course, receivers), these being known as integrated amplifiers, amplifiers are still available with separate control and power amplifier units.

The parameters of these are measured and expressed in the same way as integrated amplifiers, but each unit is measured and expressed separately. For example, input sensitivities, overload margins and S/N ratios of the control amplifier are referred to the rated output *voltage*, while the input sensitivity, overload margin and S/N ratio of the power amplifier are referred to the rated *power*.

Interfacing of these units raises no problems because they are designed specifically to partner each other. Problems may arise, however, when one make and design of control amplifier is connected to a different make of power amplifier. In this case the output voltage of the control amplifier may be too little or too much for the power amplifier.

Less well known is the fact that the impulse characteristics of the system overall can suffer due to an imprudent partnership even though the interfacing voltage and impedance are compatible.

Transient Intermodulation Distortion

One aspect of this is that for the least possibility of transient intermodulation distortion (TID) — see Chapter 8 — the designer needs to ensure that the impulse nature of the signal delivered by the control amplifier can be adequately 'digested' by the power amplifier.

This means that he has to arrange for the upper-frequency roll-off of the control amplifier to suit the *slewing rate* and negative feedback characteristics of the power amplifier. Clearly, then, if a control amplifier with a non-optimised treble-roll-off is used with a power amplifier there is the possibility that TID may occur on certain types of input signal.

Conversely, of course, there is also the possibility that the transient (music 'attack') performance of the combination will suffer unduly as the result of a too early treble roll-off. The upper-frequency bandlimiting may be applied at the input of the power amplifier or at the output of the control amplifier.

Fig. 7.5: This Sansui Model QRX-9001 receiver is equipped with a QS vario-matrix decoder in addition to decoding facilities for SQ matrix and CD-4 carrier decoding. It also has a Dolby full processor which can be used as an independent unit. Four 60W power amplifiers, each channel having logarithmically-calibrated meters for power monitoring, allow quadraphonic reproduction from virtually any source, including 'surround sound' from ordinary two-channel stereo sources.

It is certainly an invalid test to evaluate subjectively the relative performances of control amplifiers by way of a common power amplifier unless one can be assured that the impulse characteristics are fully compatible in all combinations. The same applies, of course, to the subjective evaluation of power amplifiers fed from a common control amplifier.

It will now be instructive to look at the parameters of some of the other ancillary items associated with the hi-fi system, starting with the f.m. aerial.

Aerials for F.M.

The f.m. aerial should really be regarded as a part of the front-end of the f.m. tuner section. It is in fact a tuned circuit which is coupled to the front-end through a feeder cable.

Most f.m. aerials destined for the UK are designed with a terminal impedance round 75Ω. Those manufactured for the Continent of Europe have an impedance round 240Ω, while most of those used in American countries have a terminal impedance more closely related to 300Ω.

The terminal impedance is established by the overall design of the aerial, including folding the dipole when a reflector and a director or directors are employed, because these tend to reduce the impedance of a dipole, while folding it steps up the terminal impedance again.

Impedance of Feeder

For optimum signal transfer from the aerial to the f.m. tuner the feeder cable or 'downlead' should have an impedance which matches the terminal impedance of the aerial. This is called the cable's *characteristic impedance*. Feeder for 75-ohm aerials is called *coaxial cable*, while that for 240- and 300-ohm aerials is called *twin feeder*.

It is also necessary for the aerial input impedance of the tuner to match the characteristic impedance of the feeder. Most f.m. tuners and receivers are nowadays equipped with inputs for both 75-ohm coaxial cable and 240/300-ohm twin feeder. However, if you are unlucky and your tuner has an input only for 240/300Ω feeder, you can still connect coaxial cable to it by joining the inner conductor of the cable to one terminal (leaving the other vacant) and the outer braid to an earth terminal or chassis point, as shown at (a) in Fig. 7.6.

The alternative, and sometimes better way, is to interpose a low-loss matching transformer (called a 'balun') between the 75-ohm coaxial cable and the 240/300-ohm aerial input, as shown at (b).

Attenuation of Feeder

Assuming a correct impedance match, the amount of signal lost during its travel from the aerial to the tuner depends essentially on the attenuation of the feeder. This is expressed as so many dB per given length and frequency — the longer the cable and the higher the signal frequency the greater the attenuation.

Poor quality cable may introduce as much as 6dB attenuation over a fairly long run at 95MHz. This means that if the aerial voltage (p.d.) of the wanted signal is, say, 500μV, the tuner will receive only 250μ(p.d.). Half the signal is lost in the cable.

Good quality, low-loss cable, on the other hand, may introduce no more than 2dB over the same length, in which case from the same aerial signal the tuner will receive 397μV. In general, the losses are less than this on an average feeder run.

Nevertheless, there is not much sense in spending a lot of money on a high gain aerial if a fair amount of signal is going to be lost in the feeder. If you are interested in long-distance reception trials on f.m., or if you live in a fringe area or at a location where the signal is not all that strong (particularly from stereo transmitters) then, to partner the high gain aerial (see below) you will probably be using, you should certainly invest in low-loss feeder.

Aerial Gain

On a diminishing scale, the gain of an f.m. aerial is dependent on the number of elements in its design. Gain is commonly expressed in dB relative to a dipole (single-element aerial). Thus, if the gain of an aerial is, say, 6dB over a dipole, it will 'capture' twice as much signal as a dipole in the same signal field.

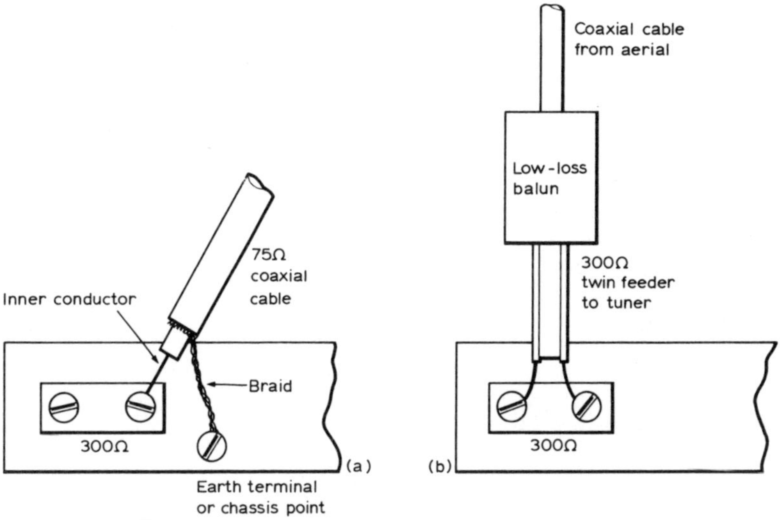

Fig. 7.6: Methods of connecting 75-ohm coaxial cable from the aerial to the 240/300-ohm aerial terminals of an f.m. tuner with the least coupling loss, (a) by connecting the braid to an earth terminal or point on the tuner chassis and the inner conductor to one terminal, leaving the other vacant, and (b) by the use of a 'balun' (standing for balance-to-unbalance) transformer. Method (b) is to be preferred.

Directionality

However, when additional elements are employed, an aerial becomes directional and it is only when such an aerial is 'beamed' to the required station that the extra gain is achieved. When the aerial is pointing away from the station then it may pick up less signal than a dipole (a vertical dipole has an omnidirectional characteristic while a horizontal dipole has a figure-of-eight characteristic). In other words, the gain is achieved only in the forward direction.

Beam Width

The beam width is expressed as so many degrees either side of the forward direction where the effective gain falls by 3dB. A multi-element aerial may have a beam width of ±30 degrees.

Front-to-Back Ratio

A well designed directional aerial will hardly respond at all to signal arriving at its rear. If we refer a dipole to 0dB, then the forward response of a directional aerial may be +6dB and the back response −30dB. The 36dB ratio between the two is called the front-to-back ratio of the aerial.

Polar Diagram

All the factors of directionality are expressed by the aerial's polar diagram, an elementary example of which is given in Fig. 7.7. Here the broken-line circle represents the response of a dipole, which is referred to 0dB, while the full-line plot is the response of a well designed six-element aerial. Relative to a dipole, therefore, this is seen to have a forward gain of 7dB.

At 30° either side of the forward direction the gain is down to 4dB (i.e. 3dB less than the forward gain), so the beam width of the aerial is ±30dB. The response at the rear is approximately −30dB, which is difficult to define

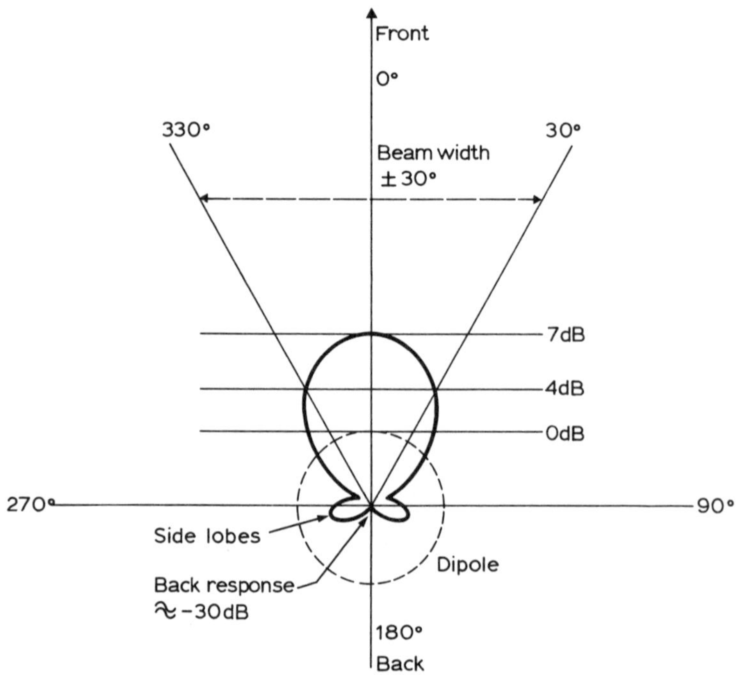

Fig. 7.7: Illustrating the primary aspects of a polar diagram (see text for details).

accurately because the curves are plotted on a linear scale. The front-to-back ratio is thus around 37dB.

It will also be seen that the response to signals arriving at the sides of the aerial is well down — apart from the small side lobes which not uncommonly appear in the polar diagrams of high gain Yagi aerials of the kind used for f.m. and TV.

The high directionality makes it possible to discriminate against strong, unwanted signals arriving from directions away from the wanted signal. For example, it may sometimes be desirable to orientate the aerial for the least response of an interfering station, even if this means losing some of the strength of the wanted station, if the tuner's front-end is being overloaded by the interfering station.

Directionality is also useful for discriminating against multipath interference (reflected signals) and thus for reducing multipath distortion (see Chapter 3 under *Capture Ratio*). A directional aerial on a rotator makes it possible to 'beam' on to distant stations, a technique used by DX buffs.

Bandwidth

This expresses the frequency range over which the aerial usefully operates. An f.m. aerial should have a bandwidth from about 88MHz to 100MHz at least — to a higher frequency if you are interested in receiving European stations. The polar diagram and the terminal impedance should not vary much over the bandwidth, which is the hallmark of good aerial design.

Headphones

Most headphones are of the moving-coil type, adopting the same principle as moving-coil loudspeakers but with a significantly reduced power input requirement. The type indicates the operating principle while also signifying other features. The two principles which predominate are moving-coil (sometimes termed *dynamic*) and electrostatic.

The moving-coil type can be plugged directly into a hi-fi amplifier, tape machine, etc. for monitoring, but the electrostatic type require the use of an additional unit which provides the polarising potential. This is generally also plugged into the mains supply, although in some cases the polarising potential is derived from signal rectification. Also, a fairly high drive *voltage* is needed for electrostatics, which is derived from a transformer in the unit.

Electrostatic headphones are not usually very suitable for connecting to a tape machine or preamplifier, and when used with a power amplifier connection cannot usually be made to the headphone jack socket. The signal for the add-on unit is commonly fed from the loudspeaker terminals of the amplifier.

Headphone Weight

This parameter can be important if protracted use of the headphone set is contemplated; you should also take account of the weight of the connecting cable! Weight including the cable averages about 350g, though there are some lighter (and heavier) ones.

Headphone Impedance

Headphones have quite a wide range of impedance. Some models are quoted at 4-8Ω, others at 16, 200, 600 and 1000Ω or more. With most of them, however, direct connection to the headphone jack socket on the amplifier, etc. is possible, except for the relatively insensitive, low impedance designs, which sometimes work better connected to the loudspeaker terminals.

Some cassette tape recorders are critical of headphone impedance and if the impedance is not just right the recording level may reduce when the headphone set is plugged in while recording. This is indicated by a fall in VU meter indication.

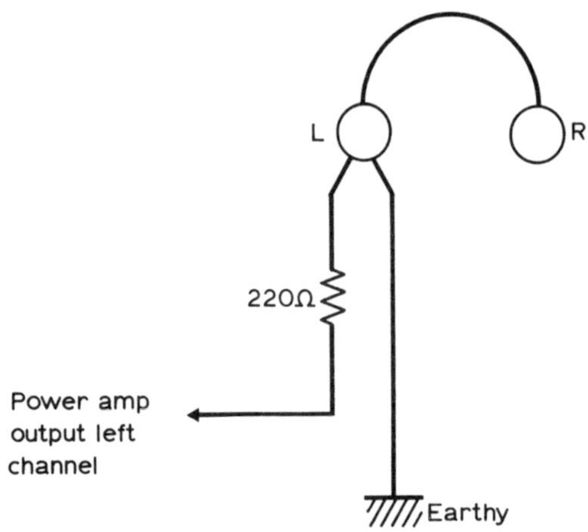

Fig. 7.8: Each unit of a headphone set is connected to the appropriate channel of the amplifier *via* a 'hold off' resistor as shown. This limits the current and hence the signal voltage across the unit.

Headphone Sensitivity

This parameter is expressed as the input voltage required for a stated SPL. For example, a parameter of '96dB SPL/100mV' indicates that each unit requires 100mV of signal voltage across it to yield 96dB of SPL.

Maximum Input

This states the maximum voltage that each headphone unit can withstand before failing. Some models can accommodate 5V r.m.s. or more with provision for 14dB transients above this. Used normally, it is unlikely that a headphone set would suffer damage as the result of over-drive. This is because each unit is connected to the power amplifier *via* a resistor of 100Ω or more (usually more like 220Ω), as shown in Fig. 7.8.

You could, of course, damage a headphone set by connecting it across the loudspeaker terminals. A 50W amplifier, for example, could produce up to 20V r.m.s. of signal when driven hard. When connected to the headphone socket, however, the series resistor limits the current and hence the voltage developed across each unit.

Frequency Response

Hi-fi headphones are capable of extremely high quality of reproduction when correctly fitted. The low bass notes require intimate acoustical coupling to the ears, while the high treble notes require a diaphragm of low effective mass. Good quality headphones have a frequency response from, at least, 20Hz to 20kHz.

Distortion

Measurements of headphones are made with a dummy ear, including SPL output relative to input voltage (e.g., sensitivity), frequency response and distortion. As with loudspeakers, the distortion tends to rise with increasing input and hence SPL; but good designs should not produce much more than 0·5 total harmonic distortion at 100dB SPL.

Cordless Headphones

In recent times, headphones have been made which operate from a mini-infrared receiver. This eliminates the connecting cable from amplifier to headphone set. At the amplifier is connected a mains-powered transmitter which incorporates light-emitting diodes (LEDs) operating in the infrared spectrum.

The diodes are modulated with the audio signal at the amplifier and the signal is thus propagated within the room as 'invisible light'. The receiver, which uses 'photo cells' responsive to the infrared emission, effectively demodulates the signal and changes it back to audio signal again. The audio is then amplified and fed to the headphone set.

CHAPTER EIGHT

DO BETTER SPECS LEAD TO BETTER SOUND?

AFTER ALL IS SAID AND DONE, a hi-fi system is finally judged by its owner (and by his friends) in terms of the realism or 'accuracy' it imparts to the reproduction. There is no doubt that to many critical ears one system or, indeed, a specific item of hi-fi equipment may well give a different listening experience from another. The problem is to determine conclusively whether the difference is for the better or worst.

There are some cases where the difference is obvious and all agree that, say, amplifier A 'sounds' better than amplifier B. With hi-fi at its *n*th degree, however, we are always dealing with extremely small, subtle differences and unless one is extremely careful judgement can be coloured by personal preferences and sometimes imagination!

Orchestra Sound

No hi-fi system yet evolved is capable of absolute accuracy of reproduction. Nevertheless, all systems worthy of the hi-fi label are capable of excellent sound quality. The sound produced by an orchestra in a concert hall undergoes an astonishingly large number of complicated processes before it arrives as sound again in the listening room. It is amazing that the accuracy of reproduction is as high as it is.

Not all hi-fi listeners are regular concert goers and most of us receive the majority of our music electronically from loudspeakers. We have established a fair impression of the 'norm' for this kind of reproduction — from small TV loudspeakers and the mini-speakers in transistor sets — so when we experience something in advance of this we are impressed. Hi-fi to most of us, then, is top quality sound reproduction. To the devotee, though, this is by no means the end of the matter.

Subtle Differences

This highly dedicated enthusiast has often trained himself to detect the most subtle of *differences* between systems and components of systems; but even for this chap to state conclusively that one item renders greater '*accuracy*' than another he must have the original live music datum established clearly in his mind. He must, therefore, be a regular concert goer and possibly music lover.

There are not many of this type of person around who are able to recall from memory the sonic impression experienced live at the concert hall. Other subjective factors may also be influencing judgement.

Comparative Audition

There is a current trend for the evaluation of hi-fi systems and parts of systems to be made by comparative audition alone. A number of amplifiers, for example, may be compared one against the other or against a previouly established reference and then judged for sonic desirability. This scheme undoubtedly has its merits; but unless handled correctly literally bristles with ambiguities.

Sonic differences there certainly will be. These are noted mostly between loudspeakers and pickup systems, including record decks; but even different amplifiers and f.m. tuners 'sound' different. Whether one 'sounds' better or worse than another is left to the judgement of the listener. If a solitary listener makes the judgement the accuracy of the results obviously cannot be very high.

The results are, then, the opinion of one person. Another person may come to a different conclusion! Apart from this danger, the expression of the differences noted presents very real problems. Since this mode of hi-fi equipment evaluation has come into being a whole new vocabulary has been evolved in an endeavour to define some of the subtle sonic differences so experienced. Unhappily, some of the expressions coined are extremely hazy in meaning and can be interpreted differently by different people. This will always be so when endeavours are made to express the performance of hi-fi equipment subjectively.

Nevertheless, a full evaluation of a hi-fi system or an item of hi-fi equipment certainly demands listening tests. The tests must be undertaken by a panel of listeners under controlled conditions. Moreover, each member of the listening panel must himself be 'tested' and weighted. Furthermore, these tests must complement — certainly not replace — the more scientific measurements undertaken in the laboratory.

'Specmanship'

A number of factors have led towards subjective attempts of evaluation. One of these is best described as 'specmanship'. That is, in recent years there has been increasing competition between equipment manufacturers and designers to achieve what is believed to be the best set of *engineering specifications.* Current technology makes it possible to design hi-fi equipment with specifications far in advance of those of the earlier valve era; but there is reason to believe that not all these advancements lead to improved sonic performance or, indeed, to greater accuracy of reproduction.

Tests undertaken by our own listening panel and also by other people have indicated that in some cases a preference is shown in favour of some of the earlier valve amplifiers when compared with certain very highly specified transistor amplifiers.

In some areas this has led to the conclusion that valve amplifiers 'sound' better than transistor amplifiers. In our judgement this is not intrinsically true. If such a transistor amplifier is re-engineered to yield parameters similar to those of the favoured valve amplifier subsequent subjective assessment is then rendered extremely difficult with judgement often going in favour of the transistor amplifier.

To secure this condition it is necessary to 'de-value' *some* of the parameters of the transistor amplifier. In other words, it appears that developing some of the parameters of transistor amplifiers to their engineering optimum, as now provided by the state of the art, the subjective experience can be impaired rather than enhanced!

Opinionated Approach

Another factor relates to the very critical nature of objectively measuring parameter differences and relating these to sonic differences. Such measurements call for highly sophisticated and hence expensive test equipment, engineers and technicians with extensive background experience, with highly developed hearing and an appreciation of live music, and a very reliable listening panel.

These requirements are tending to shift evaluation emphasis from the scientific towards the personal 'opinionated' approach. The overall cost of the latter is far less than the former, which is another factor.

After spending many thousands of pounds in the development of a new item of hi-fi equipment, it can hardly be regarded as 'fair' to the designers and manufacturer concerned if the item is given a relatively poor assessment from elementary listening tests alone.

E.S.P.

The much more scientific approach is to endeavour to discover which parameters have the greatest influence on the sound experience and how they should be regulated for the greatest accuracy of reproduction. Although this is a protracted and costly business there are now signs that this is happening in the more scientifically involved areas of hi-fi.

Our lab, too, has been researching into these aspects of hi-fi equipment design. Some of the hi-fi equipment manufacturers themselves are also investing in research along these lines. Hitachi, for example, is one company which gears electronic design to the subjective requirements. This is called 'Emotional Response Sensation and Physical Characteristics' (ESP for short!).

The scheme aims at establishing electronic design procedures based on the sonic experience which compensates for the lack of objective/subjective correlation of the more conventional electronic design approaches. The firm has evolved a way of measuring 'emotion' — translated simply to 'pleasure' and 'displeasure' — scaled to a wide range of subjects and taking account of the known pertinent parameters of a reproduced sound field.

A large number of listeners were employed, and as the various parameters were changed under controlled conditions so the reactions of the listeners were recorded and statistically analysed. The data resulting are then used at the electronic design stage, which seems to us to be a very logical way of hi-fi design.

Specs Don't Tell All

It is certainly true, then, that equipment specifications fail to tell the whole story about the performance. Indeed, the basic engineering parameters tell very little. Nevertheless, they are essential not only from the design aspect but also so that the consumer can compare one item against another in terms of technical quality and value for money.

There are, though, parameters which *do* have a close relationship to the auditioning of the equipment and it is a great pity that more manufacturers do not include these in their specifications — in addition to the basic engineering parameters.

Effect of Interfaces

It must never be forgotten that the listening impression can be greatly influenced by interfaces (see Chapter 7) and by the acoustics of the listening room.

For example, if a listening panel evaluates comparatively, say, ten different amplifiers with a given pair of loudspeakers, and then the same test is undertaken again with a different pair of loudspeakers it is possible that in some cases the second results will differ from the first. This is because some of the amplifiers may be more favourably inclined to the first pair of loudspeakers than to the second pair, and vice versa.

Similarly, a change in pickup cartridge may give a different set of results again. Some cartridges interface more happily with some amplifiers than others (see Chapter 5). Moreover, the relative characteristics or parameters of the cartridge and loudspeaker pair can influence the sonic results. Certain cartridges are notoriously bad partners for certain loudspeakers.

The interfacing of one make of preamplifier or control amplifier with a different make of power amplifier can quite significantly affect the listening experience owing to the resulting differences in impulsive characteristics (see Chapter 7). Tuner/amplifier interfacing is less critical, particularly when the tuner is designed with a relatively low output impedance.

However, differences can certainly be detected when the output impedance of the tuner is not all that low and connection is made to the amplifier *via* fairly long-length screened cables. It is also now being suggested that the nature of the connecting leads themselves can influence the final sound result*.

---

* *Jean Hiraga, Can We Hear Connecting Wires?, Hi-Fi News and Record Review, August 1977*

Effect of Acoustics

If a skilled panel evaluates comparatively a number of loudspeakers operating from a common amplifier and high quality master tape signal source one set of results will be obtained in one listening environment, while a different set of results may well be obtained in a different environment. The acoustical characteristics of the listening room can thus influence the results.

Turntable systems and record decks are affected by resonances, as discussed in Chapter 5. This is now fairly well established (e.g., see Martin Colloms, *Hi-Fi Choice — Turntables and Cartridges,* published 1977 by Sportscene Publishers Limited). However, how these resonances are likely to influence the sonic results are related to other resonances in the overall system, including those of the pickup cartridge, loudspeakers and, indeed, the listening room; also by how well the record deck is isolated from the sound field and by the intensity of the field (see under *Acoustic Feedback,* Chapter 5).

If you are in a position to make a good quality tape recording from your hi-fi and disc record source, try making such a recording (a) with the amplifer volume control turned right down (monitor on headphones if you wish) and (b) the same recording again but this time with the volume control well advanced for high intensity reproduction. If the record deck is subject to 'colouration' as the result of the sound field inciting resonances, etc. you will certainly hear this when tape recording (a) is compared with tape recording (b).

By now you will have gathered that the permutations are 'endless'. You will also begin to realise the danger of assessing on a conclusive basis the performance of an item of hi-fi hardware by listening tests alone.

Our Listening Tests

As already noted, for our assessments — linked to lab results — we employ a test panel of, at least, ten people. During the course of the tests, and unknown to the members of the panel, we deliberately switch several times to sounds with wide measurable differences between them and also to sounds with no difference at all between them (e.g., a repeat of the reference sound).

If the panel is working accurately we expect to achieve consistency of results from *each* member of the panel on each repeat. This is a sort of 'confusion' matrix which sorts out the panel members most suitable for this kind of assessment. Also, from the results obtained with different programme indices, this detailed processing leads to 'weighting' of each panel member. This, of course, is merely a rough outline of the techniques involved.

Each panel member is given a sheet comprising 50 or more (depending on the range and type of tests in hand) scales graduated over ± 5 with zero at the centre, as shown in Fig. 8.1. These avoid the use of cloudy adjectives and hazy descriptions. The plan is for each member to mark the appropriate scale as to whether the sound he (or she!) is hearing is by judgement above or below the quality and 'accuracy' of the sound previously heard as a reference. In

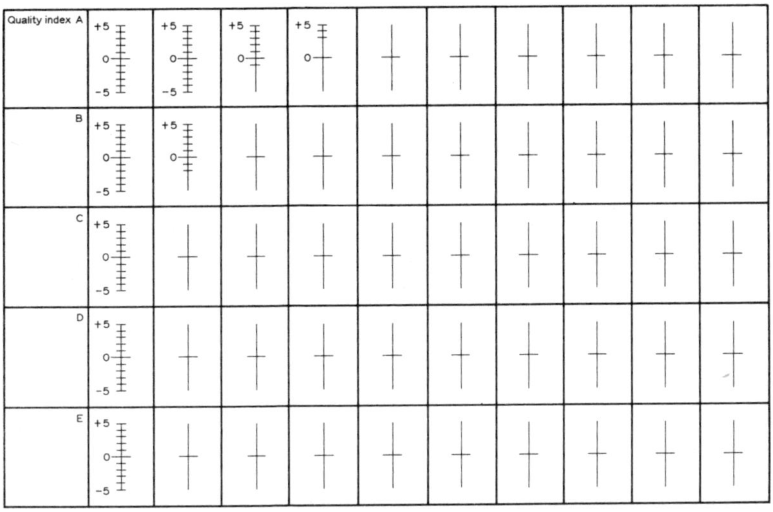

Fig. 8.1: Marking scales used by our listening panel (see text for details).

this way the panel members are subjected to the least pressure and defining brain work, which is important.

The panel is made comfortable in a room which is removed from the room housing the equipment under test, the test equipment therein and the comparative switching unit. The listening room is optimised for domestic listening reverberation time (*circa* 0·4 second average) and is acoustically insulated (avoiding spurious sounds getting in and unsociable sounds getting out!).

Sometimes a second, smaller room housing a smaller listening panel — also acoustically insulated and optimised for reverberation time — is used in parallel with the larger listening room and larger panel. This helps with the assessment of the influence of different loudspeakers on the results. A great deal of information is expanded from the 500+ results so obtained, and a computer may be used to analyse it.

Over the years we have obviously detected how variations of certain parameters can affect the listening experience, and some of these will now be looked at.

Parameters Which Influence Listening

Amplifiers with a greatly extended small-signal, upper-frequency response are commonly less favoured in a listening test than amplifiers whose −3dB roll-off point occurs round 35kHz. Also an amplifier whose treble roll-off follows a natural 6dB/octave law generally attracts a higher vote than one whose roll-off is 12dB/octave or more. On the other hand, a too-early upper-frequency roll-off tends to give low marks to a transient type music index. These things are shown in Fig. 8.2.

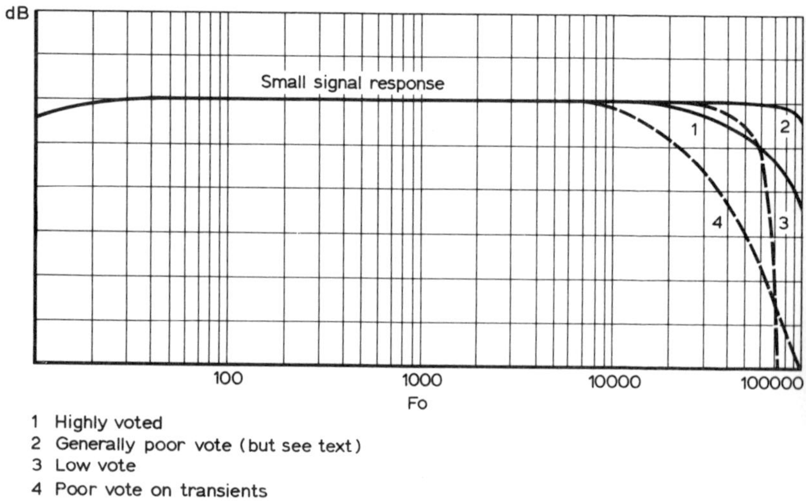

Fig. 8.2: Aspects of small-signal upper-frequency response which have been proved to influence the listening experience.

An amplifier, whose intermodulation distortion entering the *audio passband* (20Hz−20kHz) exceeds 0·5% as two equal-amplitude signals displaced by 1kHz are swept up to 100kHz with the amplifier set and signal adjusted for 1dB below peak clipping on the composite signal at middle frequencies, is commonly given lower marks than one whose intermodulation distortion under these conditions holds at 0·1 or 0·2%.

It is noteworthy that an amplifier whose small-signal upper-frequency response reaches the −3dB point at not higher than 40kHz is less likely to produce relatively high, in-band IMD than one with an extended small-signal upper-frequency response. This is a function of the slewing rate of the output transistors in the power amplifier, the effect being slewing rate induced IMD and transient intermodulation distortion (TID). More information on TID is given in our *The Audio Handbook* (Newnes-Butterworths), pages 43-44, 102 and 124-126.

The power amplifier should not be capable of being driven to slewing rate limiting as the frequency of a signal, applied to a normal input at 1dB below clipping at 1kHz, is increased up to, say, 100kHz.

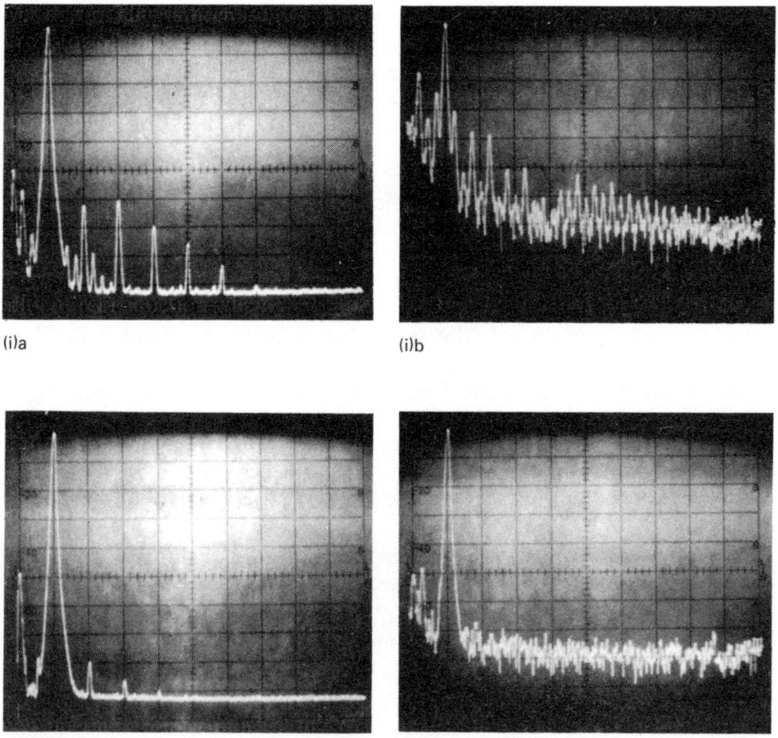

(i)a (i)b

(ii)a (ii)b

Fig. 8.3: Harmonic distortion spectrograms at 200Hz driving signal. (i)(a) one amplifier at 1dB below clipping and (i)(b) same amplifier at 40dB below clipping. (ii)(a) another amplifier at 1dB below clipping and (ii)(b) the same amplifier at 40dB below clipping. Amplifier (i) was judged by a listening panel less highly than amplifier (ii). In fact, amplifier (ii) was judged the most highly from a large number of amplifiers. This can be understood from the far 'cleaner' spectrograms of amplifier (ii) and the relative freedom from mains supply harmonics. Scale 200Hz per division horizontally and 10dB per division vertically.

Pickup Overload Margin

An amplifier whose pickup overload margin is less than 28dB at 1kHz commonly attracts fewer marks than one whose overload is 35dB or more when the programme index is from a disc record. It has also been found that slewing rate induced distortion in pickup preamplifiers seriously detracts from the overall vote given to a disc replay index, particularly when this is rich in transient material.

On some amplifiers certain cartridges, particularly those with a relatively high d.c. resistance (see Chapter 5), are given a low vote on a programme index containing bass information (also see the comments in Chapter 5 about the upper-frequency effects of pickup cartridges when loaded into certain amplifiers).

Distortion Distribution

An amplifier whose harmonic distortion distribution from a fundamental signal at 200Hz holds constant in accordance with a constant transfer characteristic at all reproducing levels from 1dB below clipping to 60dB below clipping invariably attracts higher overall marks than an amplifier whose harmonic distribution changes wildly with changing signal level. We undertook a research programme in this respect*, and the spectrograms in Fig. 8.3 illustrate some aspects of the results.

An amplifier whose mains frequency components (fundamental and harmonics up to 1kHz or more) are at a very low level of −80dB or more ref. the full output of the amplifier when the rated power of the amplifier attracts higher marks for an ambience index than one whose mains frequency components are at a higher level and extend to relatively high frequencies (see the spectrograms in Fig. 8.3).

Distortion produced by contact resistance of speaker leads etc. have a strong bearing on the results, so changeover of circuits *must* be performed so that contact resistances are eliminated; ordinary switches are unsuitable.

These, then, are some of the more subjectively correlating parameters of amplifiers. There are many others; but, unfortunately, there is just not room left in this book to detail them all — perhaps another book is called for dealing specifically with this subject!

Tuner De-emphasis

A tuner whose de-emphasis is incorrectly engineered and whose treble roll-off starts taking effect round 5kHz is always given fewer marks than a counterpart whose response is 'flat' to 10kHz. The panel can sometimes detect differences between the frequency response of tuners over 10kHz to 15kHz, depending on the type of information transmitted (our lab utilises a closed-circuit f.m. stereo transmitter of ultimate state-of-art quality taking signal from master tapes for these tests).

It must be remembered, though, that the transmitting authorities terminate the upper-frequency response at 15kHz, so there is hardly any justification for the tuner designer to maintain his response to a higher frequency.

Indeed, this can often detract from, rather than enhance, the sonic results, particularly on stereo. A tuner should really be equipped with a pilot tone and residual sub-channel filter to avoid the 19kHz tone and other higher frequency rubbish from entering the amplifier. On average, our panel has

---

* *Gordon J. King, Amplifiers — Measuring What we can Hear, Hi-Fi News and Record Review, July 1977.*

indicated that a tuner with an effective pilot tone filter auditions better than a tuner of otherwise similar characteristics but devoid of such a filter.

Some members of our panel can actually hear the pilot tone from the loudspeakers when there is no filtering at the tuner. Moreover, the pilot tone, when of relatively large amplitude (*circa* −34dB below full modulation without filtering), can incite in-band intermodulation products both in the tuner itself and in the partnering amplifier; and these are detectable under certain conditions and with certain music indexes.

Stereo Separation

A tuner whose intrinsic stereo separation averages round 40dB over the entire frequency spectrum is generally given higher marks than one whose separation is less than 25dB and whose distortion on the breakthrough signal 15% or more.

A tuner whose speaking-channel distortion rises above 1% on stereo at full modulation always achieves fewer marks than one whose stereo distortion does not reach this level until the deviation rises to ±100kHz (e.g., 33% above full modulation).

Other factors appropriate to tuners include high residual hum level, particularly modulation hum on strong signals (the panel notices this by consistently marking down on ambience indexes) and restricted and non-linear phase i.f passband.

Interface Intermodulation Distortion (IID)

Another recent finding is IID, researched by Matti Otala, Jorma Lammasniemi and colleagues, which has been mentioned on pages 36 and 39.

Harmonic Distortion

In certain cases it has been found that a small degree of even-order harmonic distortion is sometimes better subjectively than virtually zero distortion, as this can disguise the high odd-order harmonic distortion of some programme signal sources, giving a more palatable sound.

INDEX